CW01202753

Terra Australis
INCOGNITA

To Henrik Arvid Estensen

with love

Terra Australis INCOGNITA

*The Spanish Quest for the
mysterious Great South Land*

MIRIAM ESTENSEN

ALLEN&UNWIN

This project has been assisted by the Commonwealth Government through the Australia Council, its arts funding and advisory board.

Copyright © Miriam Estensen 2006
Maps by Ian Faulkner

All rights reserved. No part of this book may be reproduced or transmitted in any form or by any means, electronic or mechanical, including photocopying, recording or by any information storage and retrieval system, without prior permission in writing from the publisher. The *Australian Copyright Act 1968* (the Act) allows a maximum of one chapter or 10% of this book, whichever is the greater, to be photocopied by any educational institution for its educational purposes provided that the educational institution (or body that administers it) has given a remuneration notice to Copyright Agency Limited (CAL) under the Act.

Allen & Unwin
83 Alexander Street
Crows Nest NSW 2065
Australia
Phone: (61 2) 8425 0100
Fax: (61 2) 9906 2218
Email: info@allenandunwin.com
Web: www.allenandunwin.com

National Library of Australia
Cataloguing-in-Publication entry:

Estensen, Miriam.
  Terra Australis incognita: the Spanish quest for the
  mysterious great southern land.

  Bibliography.
  Includes index.
  ISBN 978 1 74175 054 6.

  1. Discoveries in geography, Spanish. 2. Australia –
  Discovery and exploration – Spanish. 3. Oceania –
  Discovery and exploration – Spanish. I. Title.

994.01

Typeset in 12/17 pt Bembo by Midland Typesetters, Australia
Printed in Australia by Ligare Book Printer

CONTENTS

ACKNOWLEDGMENTS

Spanish exploration of the early sixteenth century is well documented and the subject of numerous accounts, even fiction. Far less well known are the expeditions that at the end of the century and at the beginning of the 1660s explored the great wastes of the South Pacific Ocean, gradually adding to the maps of the world many of the island groups whose names are staples of tourist brochures today. Almost forgotten is the fact that the strait that separates the continent of Australia from the island of New Guinea was traversed and first recorded by Spanish navigators. Virtually unknown are their efforts to discover viable new routes across the ocean, to establish colonies and to participate in that great European endeavour, the search for the mythical southern continent of Terra Australis Incognita. This book is an effort to bring forward across four hundred years the endeavours of these men, their small triumphs and greater tragedies, their encounters with an unimagined island world and their contributions to the sum of knowledge of our planet.

Despite the lack of popular awareness of these events, many primary sources relating to them have been extensively researched and translated, making my efforts that much easier. From their translations and reconstructions of sometimes contradictory material, I have been able to extract what appears to me a likely sequence of events. I am very grateful, therefore, for the work previously done by Sir Clements Markham, Lord Amhurst of Hackney and Basil Thompson, Fr Celsus Kelly, Gerard Bushell, Henry N. Stevens and George F. Barwick.

I owe appreciation also to the many persons and institutions whose interest and assistance provided access to numerous documents and where necessary, permission to reproduce material in their possession: the Fryer Library of the University of Queensland, Brisbane; the National Library of Australia, Canberra; the Ministerio de Cultura de España, Simancas (Valladolid), Spain; The Hakluyt Society, London; and Fr John E. Keane of the Franciscan Provincial Office, Waverley, New South Wales. For time, support and assistance of many kinds I thank Jenny Gibbs and Patrick Baker of the Western Australian Maritime Museum, Fremantle; the Embassy of Spain, Canberra; the Consulate General of Spain, Melbourne; the State Library of New South Wales, Sydney, with particular thanks for the help of Jennifer Broomhead, Mark Hildebrand and Warwick Hirst and to Paul Brunton, Senior Curator of the Mitchell Library, my special appreciation of his always generous support of my work. I thank Marion Henderson of Mt Pleasant, Western Australia for her assistance on behalf of the late James Henderson, and am indebted also to Dr Nanne Sjerp of Murdoch, Western Australia. For his unfailing support in many ways I thank Dr William F. Wilson of 'Bass River' and Melbourne; and for much generous help and interest, R.H. Forsyth of Mt Isa, Queensland; Pat Smith of the Gold Coast, Queensland; Philip Simpson also of the Gold Coast; Maylene and Mark Eccleston of Canberra and Brisbane; John Wright of Brisbane and the many others who gave time and attention to my concerns. I thank my publisher, Allen & Unwin, for its continued interest in my efforts, with a special word for Rebecca Kaiser, who did so much to bring this book to fruition. To my family, once again for forebearance, encouragement and assistance of many kinds, my gratitude always.

CHRONOLOGY

1513	Balboa sees the Pacific Ocean and claims adjoining lands for Spain.
1519—1522	Magellan's expedition circumnavigates the globe; enters the Pacific 28 November 1520.
	Cortés invades and conquers Aztec Mexico.
1526—1527	Meneses discovers part of New Guinea's northwest coast.
1529	Saavedra extends exploration on New Guinea's north coast.
1537	Grijalva seeks a new route to the Moluccas; murdered by his crew; ship is wrecked.
1556	Abdication of Charles V; Philip II king of Spain.
1567—1569	Mendaña's first voyage; discovers Solomon Islands.
1569	Francisco de Toledo Viceroy of Peru; strengthens Spain's authority; crushes Inca revival.
1571	Legazpi establishes Spanish settlement in Manila.
1577—1580	Drake's voyage; enters the Pacific September 1578.
1580	Philip II of Spain becomes king of Portugal.
1586—1588	Cavendish's voyage; captures *Santa Ana* off Lower California 1587.
1594	Richard Hawkins captured; prisoner to 1602.

1595—1596	Mendaña's second voyage; Santa Cruz discovered and settlement fails; death of Mendaña; Quirós takes survivors to Manila.
1598	Death of Philip II; succession of Philip III.
1605	Portuguese expelled from Ternate and Tidore by Dutch.
	Quirós' expedition sails from Callao 21 December.
1606	La Austrialia del Espíritu Santo (in Vanuatu) named by Quirós; he claims for Spain all of Pacific region to South Pole.
	Quirós leaves expedition for New Spain (Mexico).
	Torres in Torres Strait.
	Dutch treaty with sultan at Ternate; Ternate and Tidore seized for Spain by Philippines governor Acuña.
1607	Quirós leaves for Spain.
	Torres arrives at Ternate, departs 1 May; arrives in Manila 22 May; writes to King Philip 12 July.
1608	Final evidence of Torres 6 June.
1615	Death of Quirós.

LIST OF ILLUSTRATIONS

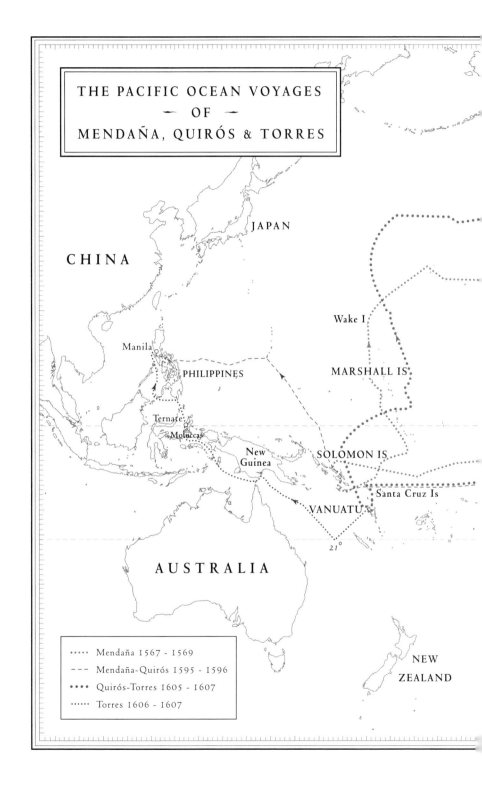

THE PACIFIC OCEAN VOYAGES
— OF —
MENDAÑA, QUIRÓS & TORRES

CHINA

JAPAN

Wake I.

Manila

PHILIPPINES

MARSHALL IS

Ternate

Moluccas

New
Guinea

SOLOMON IS

Santa Cruz Is

VANUATU

$21°$

AUSTRALIA

NEW
ZEALAND

····· Mendaña 1567 - 1569
--- Mendaña-Quirós 1595 - 1596
•••• Quirós-Torres 1605 - 1607
······ Torres 1606 - 1607

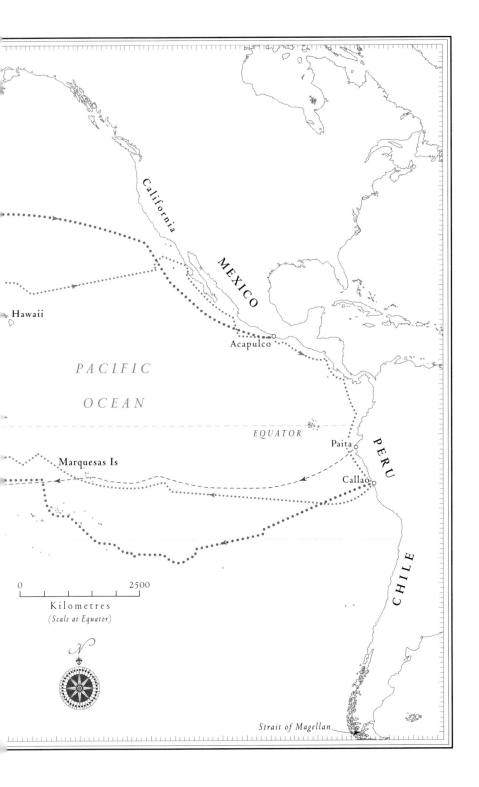

California

MEXICO

Hawaii

PACIFIC

OCEAN

EQUATOR

Acapulco

Paita

PERU

Callao

Marquesas Is

0   2500

Kilometres
(Scale at Equator)

N

CHILE

Strait of Magellan

PROLOGUE

'Two ships and one launch, in good condition and well fitted out, left the port of Callao on 21 December 1605 at four in the afternoon.'[1] So wrote the Franciscan friar, Martín de Munilla, watching from the deck of the flagship of the little fleet as it drew away from the wharves of Peru's principal port.

The expedition's commander, Pedro Fernández de Quirós, went further:

the ships being ready, the various banners were displayed from the mastheads and tops, and the royal standard was hoisted, the yards were raised, and the anchors got up in the name of the most holy Trinity. The sails were set, and the men on their knees prayed for a good voyage . . . All the artillery, muskets, and arquebuses were fired off. The ships passed near the other royal ships, which were saluting with their pieces, with many people on their decks and galleries, and many more in the town, on balconies and roofs, and on the beach . . . [2]

Thus amid the booming of gun-salutes, the fluttering of flags and banners and the shouts, cheers and prayers of the populace, the three vessels turned west-southwest as the disc of the sun touched a golden horizon and sank from view. Beyond the rim of the earth, somewhere in the far reaches of the Pacific Ocean, lay the object of their search, the great, mysterious South Land, Terra Australis Incognita.

The expedition would record islands hitherto unknown to Europeans and would emerge having made a significant geographical discovery relating to the unsuspected continent of Australia, a discovery that would lie hidden in Spanish archives for centuries. The journey which began on that December day was the last of Spain's great maritime explorations, and the end of a story, not the beginning.

In the late 15th and early 16th centuries the Spanish kingdom of Castile was drawing together the Iberian peninsula's disparate regions under a single rulership and tentatively extending maritime exploration westward into the Atlantic Ocean. In 1479 Spanish claims to the Canary Islands were recognised by Portugal, and in the quarter century that followed Spain effectively conquered the archipelago. However, active Spanish interest in what lay beyond the great western Ocean Sea had to follow the 1492 conquest of the Islamic kingdom of Granada, when the Spanish monarchs, Isabel and Ferdinand, were finally in a position to channel resources towards overseas exploration. Christopher Columbus appeared on the scene and Spain's new direction was firmed with his landfall in the West Indies in October 1492. Conflict with the territorial aspirations of Portugal now resulted in a series of papal bulls promulgated in 1493 by Pope Alexander VI, which established an imaginary line from pole to pole 100 leagues west of the Atlantic Ocean's Cape Verde Islands. Spain was to receive exclusive rights to the territory west of the line, and Portugal to maintain her claims to the east. Christian lands were to be exempt in either case. Papal diplomatic intervention in international disputes was not new. Popes had traditionally acted as mediators between Christian nations and also were accepted by Christian authorities as having special control over relations between Christians and pagans.

In June 1494, by the Treaty of Tordesillas, Spain and Portugal moved the papal line to 370 leagues west of the Cape Verde Islands, by today's measurement between 48° and 49° West of Greenwich. Thus Spain secured dominion over what would become the Americas, with the exclusion, as it developed, of a section of the South American east coast in today's Brazil. The problem of extending the Tordesillas Line onto the opposite side of the globe remained. The difficulty of even approximating longitude was compounded by several different estimates of the circumference of the earth and the varying values in Europe for the unit of measurement employed, the league. This placed the Spice Islands of the Moluccas, modern Indonesia's Maluku, in an indeterminate position subject to the claims and counterclaims of Spain and Portugal.

Spanish occupation of the Caribbean area widened, and within 30 years of Columbus' first landing, Spain had laid claim to the entire sweep of the West Indian islands. In 1513 Vasco Nuñez de Balboa crossed the Panamanian isthmus and saw before him the Pacific Ocean. A few days later, wading into the surf, he claimed possession of the Mar del Sur, the Southern Sea, and its adjacent lands for the Spanish king. In 1519 Hernando Cortés led some 600 Spanish soldiers and sailors into Mexico and established Spain's authority from the Caribbean to the Pacific. Thirteen years later Francisco Pizarro landed a force of 183 men on the coast of Peru. Inca resistance was overwhelmed and by 1535 a Spanish city, Lima, had been founded, and in two years its port, Callao. By some 45 years later Spain had established a stable viceregal administration. Fronting Spain's long, virtually uncontested Mexican and South American coastline, the Pacific Ocean was now the magnet for further discovery.

Spanish exploration of the Pacific had begun in 1521 when three surviving vessels of Ferdinand Magellan's squadron of five small ships emerged into the world's largest ocean from the strait that now bears his name. The four-month crossing of the vast ocean was one of horrendous suffering and remains one of history's greatest maritime achievements. After three years, one ship and eighteen Europeans returned to Spain. They had accomplished the first recorded circumnavigation of the planet.

Despite its terrible losses, the voyage inspired further exploration of the Pacific. In 1525 Charles, King of Spain and Holy Roman Emperor, despatched a flotilla of seven well-equipped ships, an expedition which disintegrated in catastrophe—shipwreck, desertion and disappearance at sea. Eleven years later the voyage of Hernando de Grijalva ended in mutiny, Grijalva's murder and finally shipwreck off New Guinea. New Guinea's north coast had been seen some ten years earlier by the Portuguese navigator Jorge de Meneses and in 1529 by the Spanish explorer Alvaro de Saavedra Cerón, but its south coast remained unknown. Expeditions to conquer and colonise the Philippines left Mexico in 1542 and in 1564. A permanent Spanish presence in the islands was established, chiefly at Manila at 14° 35' North latitude, but various reports in the mid-1500s of islands lying below the Equator in the seas west of South America failed to sufficiently interest government authorities. Nevertheless, the fact of an unexplored southern region seemed to confirm the belief that in that vast unknown space there must lie the huge hemispheric South Land, the mysterious Terra Australis.

Between 1567 and 1605 the idea suddenly gained prominence, and a new round of Spanish maritime exploration into the Pacific came vigorously to life. Three expeditions sailed from Peru's port of Callao, each despatched with an extraordinary purpose: the discovery and colonisation of Terra Australis Incognita.

SPE ET METV.

Chapter One
GOLD, SOULS AND
A MYTHICAL CONTINENT

The story begins with a question: what drew the 16th century navigators of Spain onto the world's largest ocean, where small, clumsy wooden vessels with relatively primitive methods of navigation faced enormous uncharted distances, unknown weather conditions, hostile natives and the hazards created by their own internecine quarrels? On 5 April 1529, by the Treaty of Zaragoza, Charles of Spain had ceded to Portugal his claims to the spice islands of the Moluccas for the considerable sum of 350 000 ducats. Underlying the agreement was a provisional demarcation line at 297.5 leagues east of the Moluccas. This was assumed to be the location of the countermedian of the line through the Atlantic which the two nations had agreed separated their halves of the world. Spanish ships were not to penetrate west of this line.

Thus Spain was denied the opportunity for wealth from the Spice Islands. Mexico and large areas of South America had been explored, leaving little opportunity for new discoveries. At all levels, however, the incredible wealth of Spain's American conquests, the gold and silver that flowed from the mines of Mexico, Peru, Chile and today's Columbia, stirred the conviction that there must be lands, probably to the west, that would yield further treasure. Spain's territorial claim to some undiscovered continent, with untold riches to be reaped by the Spanish crown and by Spanish colonists, became a mesmerising idea.

The concept of a vast southern continent went back 25 centuries to the deductions of the ancient Greeks. Pythagoras and his school had reasoned that as the sun and the moon were obviously spherical, logic dictated that the earth must be so as well. Further, the sphere was the perfect form. Fully aware of the effect of perspective, the Greeks understood that a curved line that was long enough could appear to be straight, hence the deceptively flat appearance of the earth. In his book *On the Heavens*, Aristotle wrote of the earth, 'its shape must necessarily be spherical'.[1] A century later Eratosthenes calculated the circumference of the globe.

Logic went further. The world known to the Greeks stretched across the earth's northern hemisphere, and to maintain the planet's stability in space, there had to be a corresponding terrestrial mass occupying its southern half. The idea was virtually lost during the long hiatus of the Middle Ages, when it was believed that an equatorial belt so scorched by the sun that it could not be crossed must lie to the south of the known world. The inhabitants of any region so cut off could not be descended from accepted biblical origins, and it was therefore thought that such a place could not exist.

In the early Renaissance, however, the idea of a vast antipodean continent was rediscovered by European scholars in the writings of the second century astronomer and geographer, Ptolemy. A map derived from his *Guide to Geography* and printed in 1482 depicted a southern continent linked to Africa and to Asia. Yet within 40 years the far-ranging voyages of Bartolomeu Diáz, Vasco da Gama and Ferdinand Magellan had shown that the unseen south land had to be separated by sea from the known continents. In 1538 the great cartographer Gerardus Mercator published his first world map, *Orbis Imago*, on which he sought to incorporate all the most recent geographical information. On a great southern continent he inscribed, 'That land lies here is certain but its size and extent are unknown.'[2] In 1570 Abraham Ortelius published *Typus Orbis Terrarum*, which similarly depicted a huge continent virtually enveloping the earth's southern hemisphere. It was confidently entitled Terra Australis Nundum Cognita, 'Southern Land Not Yet Known'. Some cartographers, like Martin Waldseemüller, were more cautious and showed only open sea to the south, but most were convinced that an immense land mass occupied the unexplored southern vastitude of the planet.

Beliefs in faraway lands of great wealth had existed for centuries. Classical geographers of Greece and Rome wrote of Khryse, a golden island or later, a golden peninsula. The location of biblical Ophir, from which ships had brought cargoes of gold for the adornment of King Solomon's temple at Jerusalem, was at various times thought to be in Arabia or on the African east coast. However, as trade and exploration widened, speculation moved these supposed fountainheads of treasure farther and farther away. By the 16th century Ophir was linked to the

Pacific. The fact that the principal Spanish transpacific voyages tended to remain north of the Equator added to the mystery of the southern seas. In the middle of the 16th century a series of world maps appeared, produced in the French city of Dieppe, which depicted a continental land mass named Java la Grande occupying very roughly the region of modern Australia. Fancifully drawn, they might have encouraged further the concept of a large exotic country to the south. The belief that such a continent lay somewhere in the higher southern latitudes would not be wholly dispelled until the discoveries of James Cook on his second voyage of 1772–75 were made known.

In Peru there circulated rumours of mysterious islands, rumours of both Spanish and Inca origin. In a report to Philip II in 1565, the interim governor, Lope García de Castro, discussed the proposal of a merchant, Pedro de Ahedo, for a voyage to some islands somewhat confusingly 'called Solomon, which lie opposite to Chile and in the region of the Spice Islands . . .'.[3] The discovery of several small islands off the Chilean coast by the navigator Juan Fernández further fuelled rumours that Terra Australis was not far away. Spanish imaginings of this nature were, however, flawed in a way they did not realise. In crossing the Pacific in 1520–21, Magellan's ships had been carried on currents setting in the same direction, from east to west, and the explorer had underestimated the distance he covered. This was compounded by the fact that longitude could only be approximated by a procedure many navigators ignored. Pilots sailing from the Americas thus greatly misunderstood the breadth of the Pacific.

Spain's plans for Pacific exploration entailed a powerful religious motive. Catholic Spain had fought for centuries against the tide of Islamic invasion, in Spain and elsewhere, and many Spaniards saw themselves as the champions of Christendom, crusaders morally charged with bringing Catholicism to the non-Christian world. The religious convictions of the Spaniards of the 16th century held a powerful place in their daily activities, and a similarly strong influence was exerted by religion on the political outlook of their rulers. The importance of bringing the faith to indigenous peoples was never doubted.

In the Americas, Spain's spiritual obligations were enhanced by the institution of the Patronato Real de las Indias, literally Royal Patronage of the Indies, a large body of exclusive powers in handling ecclesiastical affairs granted by Pope Julius II to Spanish kings in 1501 and 1508. This laid upon the king and his *conquistadores* a further responsibility in promoting the Catholic faith among the people they conquered by providing and supporting missionaries. From this unique arrangement there was drawn the premise that Spain's chief mission was evangelical, that the nation bore an obligation to seek and convert the pagan world. It was a responsibility taken seriously by the religious orders and many of Spain's leaders. Expeditions were almost invariably accompanied by friars, although the average soldier or sailor heading into the unknown probably saw the glitter of gold as his primary motivation and service to God a necessary second.

Beyond these incentives hovered the mirage of the Great South Land. The eminent Spanish scholar Juan Luis Arias de Loyola imaginatively wrote to his king, Philip III, that the land was manifestly 'as fertile and habitable as the northern hemisphere'[4] and 'greatly stored with metals and rich in precious stones and pearls, fruits and animals'.[5] Of the immense size of this unseen continent there was no question. On the reverse of a 1593 map by the cartographer Cornelius de Jode, which depicted the north coast of New Guinea, there appeared the statement: 'On the south of this region is the great tract of the Austral land, which when explored may form a fifth part of the world, so wide and vast is it thought to be'.[6]

A land so immense was assumed to be well inhabited, and it was for this population that philosophical reasoning, supported by Scripture, demanded conversion to Christianity. By the 1580s the spread of the Reformation had given further impetus to Catholic motivation. Territorial ambitions were compounded by the growing urgency of saving souls.

Insofar as it was known, the Pacific Ocean was regarded by Spain as its own. Its great waters, stretching from the Philippines to the Americas, were immediately accessible from the ports of Spain's American colonies, but remote from the maritime reach of other European nations. Its approach from the east was guarded by the treacherous passage and subantarctic weather of the Strait of Magellan. On its western edge the Portuguese, after 1529 left in virtual possession of the lucrative spice trade, had little reason to push eastward. Nor did the Asian countries display particular interest in the far ocean wastes. Thus for some five decades the Pacific, virtually inviolate, was the 'Spanish Sea'.

SPE ET METV.

Chapter Two
THE FIRST VOYAGE:
1567–1569

1589

B y the mid-1500s two Spanish viceroyalties occupied the long American coastline facing the Pacific Ocean. One was Nueva España, or New Spain, essentially today's Mexico and Central America. The other, Peru, incorporated all of South America except Venezuela, which was part of Nueva España, and Portuguese Brazil. Each was administratively divided into a number of *audiencias*, or captain-generalcies. Peru derived great riches mainly from the region itself, from the fabulous silver mines at Potosí, Huancavelica's deposits of mercury, essential in the processing of silver, and from gold mines in today's Columbia. Control of the wealth and power of this huge territory lay in the viceregal city of Lima, founded by Francisco Pizarro on 6 January 1535, the feast of the Epiphany. Thus it was also known as Ciudad de los Reyes, City of the Kings. Laid out in square blocks, it sported fine buildings and an elaborate court life. Roads linked it to its port, the natural harbour of Callao, some 13 kilometres to the west.

In about 1557 a Spanish soldier named Pedro Sarmiento de Gamboa arrived in Peru. Perhaps born in 1532, Sarmiento joined the army at eighteen, and fought for five years in Spain's European wars before leaving for the Americas. Reputedly a mathematician and geographer, he might also have acquired some navigational skills during the transatlantic voyage. Sarmiento spent perhaps two years in Mexico and Guatemala before suddenly and mysteriously fleeing to Peru. Clever, ambitious and ruthless, he apparently ingratiated himself into the viceregal courts of García Hurtado de Mendoza, Marquis of Cañete, and his successor Diego López de Zúñiga y Velasco, Conde de Nieva, until Nieva's assassination. The position of interim governor was assumed in 1564 by the Licenciate Lope García de Castro.

Sarmiento developed an interest in Peruvian folklore, particularly the story of the emperor Tupac (or Topa) Inca Yupanqui, who according to tradition returned from a year's voyage across the western ocean with gold, slaves and other valuable trophies. The tale may have had some factual basis, for pre-Spanish Peruvians evidently built seagoing rafts,[1] and fragments of Inca pottery have been found on some Galapagos islands.[2]

Sarmiento also theorised that Ophir lay somewhere in the Pacific. In 1564 he survived a brush with the Inquisition, and three years later was pressuring the governor to send an expedition into the Pacific in search of treasure, either on the unknown continent or on islands along its littoral. A persuasive opportunist, Sarmiento evidently based his argument largely on the popular belief that the islands were at no great distance from Peru.

Governor Castro apparently had an additional incentive for acceding to Sarmiento's plan. Documents indicate that there were restless and disruptive elements in the Spanish colonial population. To maintain peace and order, governing authorities had developed a policy of encouraging 'idlers' and 'vagabonds' to join journeys of exploration.[3] Here Castro saw such an opportunity. To men generally drawn from the poorest levels of society, even a seaman's exiguous pay and basic daily rations could be attractive. And there was the lure of possible wealth and for some perhaps a sense of self-worth in sharing in a significant, even perilous enterprise.

The fitting-out of two ships is recorded in surviving documents. They were the *Los Reyes*, of 200 tonnes burden and designated the flagship or *capitana*,[4] and the *Todos Los Santos*, of 140 tonnes, the secondary ship or *almiranta*. Both are referred to as *naos*, galleon-type vessels, typically three-masted, square-rigged with topsails and crow's nests on the forward masts, lateen-rigged mizzens and spritsails fitted to the bowsprits. Sarmiento called them ships-of-war, but they were, in fact, merchant vessels armed with culverins provided with round- and grape-shot, arquebuses with match-cord and, as recorded, three leather bags of gunpowder. Halberds, swords and shields were provided for the soldiers. The two vessels were purchased for 10 500 pesos in silver coins on 17 June 1565, and refitting for the voyage apparently began on 1 July.

To the angry disappointment of Sarmiento, the command of the expedition went to the governor's young nephew, Alvaro de Mendaña de Neira (or Neyra), who was named captain-general 'of the voyage of discovery to the islands and mainland of the South Sea'.[5] Whether this reflected distrust of the ambitious Sarmiento is hard to judge, but there had been rumours of the planned seizure of one of the ships by would-be corsairs, and writing to the Council of the Indies in Spain the governor explained his choice of commander: a relative whom he knew to be 'free of all intrigues'.[6]

Born in Spain, probably in 1542, Mendaña had come to Peru at seventeen, most likely in the entourage of his uncle. At about age 25 and apparently inexperienced, he was entrusted with the responsibility of an expedition for which there were the highest expectations. His conduct during the ensuing voyage does not suggest effective leadership. Faced with opposition from his men, Mendaña tended to give way. This, however, could at times have been a matter of common sense. The explorer Hernando de Grijalva, also crossing the Pacific, had been murdered by his crew some 30 years before.

Mendaña's moral concerns with his mission are evident. Unsure of how to deal with native people, he consulted the expedition's friars, and his actions were generally benevolent. Clearly he empathised with the sufferings of his own men, noting later that recalling their deaths still 'touches me to the soul and overcomes me'.[7] His dedication to the conversion of heathens to the Catholic faith was sincere if, as it turned out, ineffectual. Sarmiento's mind, on the other hand, was undoubtedly much more on prospects of wealth.

Sarmiento's position on board the *Los Reyes* is not clear, but certainly it was not as important as his own writings suggest. Navigation was in the hands of the expedition's chief pilot, the experienced Hernando Gallego, born in the Atlantic port of

La Coruña, at sea as a boy and fifteen years later a qualified pilot, serving in Spain's naval campaigns in the Mediterranean. By 1557 he was on South America's Pacific coast. Skilled, practical, stubborn and efficient, Gallego knew and exerted his rights as the chief of the expedition's four pilots, the navigator to whom even the commander must defer. Conceding authority to Mendaña and navigation to a strong-minded pilot, Sarmiento, simply mentioned in the records as the expedition's cosmographer, found himself without the leadership role he evidently expected. The hostility that developed between Sarmiento and Mendaña in the succeeding months later led to years of spiteful activity on Sarmiento's part.

Chief purser of the expedition was Gómez Hernández de Catoira, who as official chronicler would write the longest and in many ways the most apparently accurate of the several narratives of the voyage.[8] Recounting the activities of others, Catoira includes little that throws light upon himself, but occasionally pauses for a thoughtful comment on events and on his fellow adventurers. On board as well was Hernando Henríquez (Fernando Enríquez), in the capacity of *alferez-general* (loosely, chief standard-bearer) and the only person on the voyage accorded the title of Don. The background that entitled him to this is not clear, but Henríquez lived by a chivalric sense of honour which won him the respect of those who served with him. With the island people the expedition encountered, he would endeavour to be fair and compassionate even in very difficult situations, with no lack of courage when circumstances required it.

The secondary ship or *almiranta*, *Todos los Santos*, was commanded by Pedro de Ortega Valencia, formerly *aguacil mayor*, or high constable, of Panama, and now the expedition's camp-master or military commander as well as ship captain. Ortega brought with him five servants who would also serve as soldiers, provisions for them and for himself, and a supply of arms and ammunition. Evidently he was a respected and competent officer, maintaining discipline among the soldiers, a potentially unruly group, and later, like Hernández, attempting to deal with native islanders in an essentially humane manner under almost impossible circumstances.

A huge variety of equipment was purchased for the expedition, for which documents record payment—flags and pennants, a forge and other necessaries for five blacksmiths; twelve pitchers of wine for each of four pilots, arms and ammunition, lead and moulds for making round shot, gun pieces and armour; stocks and fetters for troublemakers. Trade goods were taken on board. Small bells, beads and red caps are mentioned. A list of items carried by Magellan's ships in 1519–22 includes lengths of fabric, combs, fish hooks, 4800 cheap knives and 4000 brass and copper bracelets.[9] Mendaña's trade goods would have included similar items. There were also fittings for the construction of a brigantine for coastal exploration. Provisions would have included salted meat and hard-baked biscuits, normally a year's supply.

Figures cited for the ships' complements vary from 100 to 160,[10] including ship's boys, several black slaves, and 70 soldiers by one account, 60 by another. Thousands of pesos in silver are recorded for the wages of officers and seamen, mainly for six months in advance. Apparently no advance payments were made to the soldiers, whose remuneration was to come from the riches of the lands they conquered. Evidently reflecting the government's policy of encouraging social undesirables to take part in voyages of exploration, one document states that almost all the soldiers were criminals, one under sentence of death.[11] There were also two miners, experienced in prospecting for gold, and four Franciscan friars, fully equipped with altars, vestments and the requisites for the celebration of the mass and administration of the sacraments, together with a statue of the Virgin.

On 19 November 1567, the feast day of Saint Isabel, the two ships left the wharves at Callao, but in contrary winds it was the next day before they could actually sail out of the port. They steered southwest into an unknown ocean that the men on board nevertheless believed was indisputably Spain's to explore and exploit.

What were Mendaña's intentions for this voyage into the vastness of the Pacific in search of a continent that might not exist? His orders from the governor have not survived, but it appears that the initial objective was to find 'certain' islands, probably those envisaged by Sarmiento to be somewhere between 14° and 23° South latitude and no more than about 600 leagues, some 1500 kilometres, from Peru.[12] As the islands were seen as fringing Terra Australis, the great unknown continent itself would be found soon afterwards and a Spanish colony established. Some documents state that New Guinea was also an objective. The Spaniards' concept of the size of the Pacific was painfully short of reality.

For over three weeks, covering roughly 20 to 35 leagues a day, the chief pilot guided the ships west along a parallel close to 15° South latitude. There was no accurate means of determining longitude. A pilot approximated the distance in leagues sailed east or west of the point of departure, making allowances as best he could for the steering capacity of the ship, variations in wind, currents, calms and swells, even the skill of the helmsman. Spanish ships of the period did not use the log and log-line then employed by the English. According to one source, they used simply 'eye and judgment', measuring time by the sandglass. From this only a very tentative estimate of longitude could be made, which generally was not even attempted.

By 16 December, Gallego's calculations indicated that he had sailed 620 leagues 'more or less'. The sea remained empty of any sight of land. Still heading west, Gallego laid a course on just over 6° South latitude. Sarmiento was furious, and even more so when Gallego, backed by Mendaña, refused to investigate a cloud bank which Sarmiento thought could indicate one of the Inca islands.

The weather deteriorated, the ships plunging and rolling in heavy rain. Then calm returned, with light winds and a quiet sea. In the evening of the first day of the new year, 1568, one of the boys, playing with other lads, tumbled overboard. Gallego, on deck with Mendaña, saw him fall and gave the alarm, 'Onbre a la mar!'[13] Catoira recorded:

> we saw him drop astern shrieking out; we threw him out ropes but he could not lay hold of them. We threw out a hencoop, but he was not able to take hold of it on account of the darkness; fortunately for him there was not much wind, although the ship was still moving. Hernando Gallego tried to turn back for him, but . . . the ship would not go about . . .[14]

At this point there was a near collision as in the darkness the *almiranta* slid by. Now there was no answer to their shouts and 'we commended him to God; but when we hailed him again he answered, but so faintly . . .'.[15] Two sailors leapt into the sea. A 'hatch' (hatchcover) on a line was thrown after them, and shouting as they swam with it, the men reached the exhausted lad. They 'put him on the hatch' and 'drew him to the ship'. Mendaña wrote, 'It is my belief that Our Lady delivered him miraculously, because we recommended him to Her'. Some, including the boy himself, had seen a light 'like a candle' shining above him. A few days later a sailor named Juárez Méndez, or Juan Rodríguez Méndez,[16] fell overboard. On his grabbing a rope hanging from the stern he was pulled aboard, although the ship was then moving in a fair wind.

They continued west 'in the latitude of 6½ degrees . . . south of the Equinoctial', as Gallego recorded,[17] generally in fair winds, but among the crew fear was growing that they would never find land. The soldiers were bored and restless; aware of mutinous mutterings, Ortega removed their weapons and locked them away.

On the 57th day of the voyage a boy at the maintop sighted land, and by sunset the ships were approaching a low flat islet tufted with palm trees and ringed by reefs. This was probably Nui in Tuvalu, which the Spanish named Isla de Jesus and which Gallego, again not making sufficient allowance for favourable currents, calculated to be 1450 leagues from Lima, a considerable underestimate. Seven small canoes with a man in each swept out from the beach, but on reaching the ships turned back to the shore. There some sort of white material was displayed, apparently in response to the white cloths waved by the Spaniards, and a series of fires were lit and kept burning through the night. It seemed to be an attempt at communication, but what was the message? A welcome? A warning? The Europeans could only speculate. At daybreak the wind rose and rain surged in, the ships falling to leeward of the island. Despite the eagerness of the crew to go ashore, the need for fresh water and Mendaña's great desire to formally claim this first landfall for his king, beating back to the island against strong winds and currents was now too dangerous.

Fifteen days later, in the darkness of about two o'clock in the morning of 1 February, lookouts saw the pale glimmer of the sea breaking on a line of reefs. Sighted the day before Candlemas, the reef was named Los Bajos de la Candelaria, but whether this was today's Ontong Java atoll at approximately 5° 20' South, 159° 30' East, or Roncador Reef at about 6° 13' South, 159° 22' East, is argued by modern historians. For three days, in rain and strong winds, the vessels cruised along the reef, hoping to see land. None was sighted and they moved on.

Chapter Three
LAS YSLAS DE SALOMÓN

15 89

J ust before dawn on Saturday 7 January 1568, the man at the maintop sighted land at a distance of about 15 leagues. As the dawn light fanned across the ocean, the men on deck saw a long, steep-sided hump of land that had to be the continent. The ships' companies knelt in fervent prayers of thanksgiving, singing the 'Te Deum Laudamus'. By late afternoon the vessels had closed upon a mountainous, jungle-clad country with a shoreline of mangrove swamps and sand beaches. They had reached the island group that came to be known as Las Yslas de Salomón.

Mendaña's account describes native warriors, whom he called *yndios*, paddling swiftly out to the ships in small canoes 'all ready for war' with bows and arrows, spears and apparently wooden clubs.[1] Signs of peace were made, however, the natives calling for what the Spaniards understood to be a leader, the *tauriqui*. Mendaña was identified as such, and as he raised his hand and made the sign of the cross some of the natives did the same. He threw them a red cap, which they put on the head of one of the men. There was a clamour for more caps, and Mendaña tossed out three or four, beckoning the men to come on board. This, however, they refused to do until a Spanish sailor jumped into the water and swam out to get into one of the canoes. Then, rather fearfully, some 24 of them climbed up the ropes on the ship's side. The warriors were naked and very dark, the chiefs with painted faces and plumed headgear. The Spaniards embraced them and offered food and drink—wine and biscuit were rejected, but meat and preserves were consumed. The explorers distributed coloured beads, red caps, bells and other small gifts and were delighted when one of their visitors repeated after them the Lord's Prayer and the Credo, pronouncing the Spanish very well. Mendaña gave the man a shirt. Less welcome was the theft of objects which the natives flung overboard to be retrieved by those still in the canoes. Mendaña offered one man wine in a silver cup. He refused the wine but kept the cup. Another had to be prevented from seizing a ship's bell. One man climbed to the maintop with such speed that Mendaña remarked that he would make an excellent sailor. In the meantime a boat carrying soldiers had been despatched to find a suitable anchorage, returning unsuccessful at nightfall as the islanders were leaving. The ships moved seaward and tacked on and off through an anxious night.

By daylight the search for an anchorage resumed. With a man at the foretop and another on the bowsprit, Gallego piloted them over a reef into a large sheltered bay. In doing so, they saw to their wonderment 'a star shining brilliantly over the maintop'. This was probably the planet Venus, sometimes visible by day, which they took for a guide sent by God, the Virgin and the 'three Magi kings'.[2] By late afternoon the ships had dropped anchor in 12 fathoms of clear, green water in what they named Bahía de la Estrella, Bay of the Star. Mendaña disembarked with his officers, the friars and some soldiers. The vicar, Fray Francisco de Gálvez, carried ashore a large cross and held it upright as prayers were said. With the cross set up, the friars sang the hymn 'Vexilla Regis Prodeunt' (The Banners of the King Press On),[3] and Mendaña took possession of the land in the name of King Philip II. Having left Peru on the feast day of Santa Isabel and crossed the ocean under her patronage, they named the country for her. But had they reached the great southern mainland or merely an adjacent island? This remained to be discovered.

Initially the relationship between the newcomers and the islanders was amicable. Gifts were exchanged—coconuts and a white bone armlet for a red cap—and on hearing the sounds of a tabor and a fife on one occasion the warriors on board broke into a dance. When they displayed their own instruments, small pan pipes and a large conch shell, Mendaña sent for a trumpet and then had some of the soldiers sing to a guitar, whereupon there was more dancing. The two leaders, Mendaña and the chief Bilebanarra (or Vylevanarra), exchanged names in a gesture of friendship. Each group attempted to learn something of the language of the other and Mendaña took upon himself to win the chief's allegiance to King Philip and to Christianity. Of Bilebanarra he wrote:

I endeavoured to make him understand that I was a vassal of
your Majesty, and by your command had come to that country . . . to bring them the knowledge of God and of our holy
Catholic faith . . . when he asked me where the King of Castille
was, I replied he was in Castille, his own land . . . to make him
better understand . . . I took a sea chart, and showed him what
was sea and what was land, and, pointing to a very small island,
I said that this was his country, and that all the rest belonged
to your Majesty. Then he asked me where God was, and if he
was a great lord.[4]

Mendaña replied that God had made the heavens, earth, sea
and 'us all', and on Bilebanarra's repeating the lesson with signs
and words of his own, continued:

I told him that since he was my friend and brother, and that
I was a vassal of your Majesty, he ought to be one also. He
assented . . . and thereupon I caused a deed to be drawn up
in proof that he had rendered allegiance to your Majesty, and
accounted himself your vassal.[5]

What Bilebanarra understood of this document can only be
imagined.

The Spaniards continued to live aboard their ships, but came
onto the wide sandy beach to celebrate mass. They realised
that the land was heavily populated; people appeared in crowds
of several hundred. There were numerous separate territories,
some at war with each other and many, the Spaniards were told,
fiercely opposed to the visitors. Within a few days Bilebanarra's
own amiability had cooled. He received them warily, 'more in

fear than in friendship'. Gifts of food dwindled. In all probability the people were confused as to what kind of being these intruders were. When some of the soldiers went aside to urinate, they were followed and observed by both men and women. Returning to a village they had visited before, the Spaniards found it deserted.

Bilebanarra had warned the explorers that eleven or twelve chiefs were combining to kill and eat them. That was probably the Europeans' first realisation that they were among cannibals. This became all the more real one day when the Spaniards were at mass on the beach and a fleet of fourteen canoes arrived with an armed deputation from another chief, bearing a gift for Mendaña. The gift was, as Gallego recorded, 'a quarter of a boy, with the arm and hand', along with some yams, which they urged Mendaña to eat. Some of the shocked soldiers prepared to open fire, but Mendaña restrained them. Later he wrote:

> I accepted the present, and, being greatly grieved that there should be this pernicious custom in that country, and that they should suppose that we ate it . . . I caused a grave to be dug at the water's edge, and had the quarter buried in his [the native leader's] presence . . . seeing that we set no value on the present, they all bent down over their canoes like men vexed or offended, and put off and withdrew with their heads bent down.[6]

Setting their servants to felling trees and sawing planks, Gallego and the ships' carpenters began the construction of a 4- to 5-tonne brigantine in order to explore the narrow straits and reef-bound islands around them. Mendaña had to discover whether Santa Isabel was an island or, as he hoped, part of a mainland. Another

concern was food. Although the ships were still reasonably well
provisioned, these supplies had to be husbanded for the explorations before them, and the local people had now ceased to bring
anything edible. Mendaña conferred with the Franciscan vicar on
how to deal with this. Fray Gálvez said that as friendship had been
established with the local chief, 'we might very well go inland
and ask for food, paying for it with other things'. If none was
given it could be taken in moderation, but not enough 'to leave
them despoiled, nor take any of their goods, nor their women
nor children'. If they 'broke the peace' the Spaniards could only
defend themselves but keep the food.[7]

No effort seems to have been made to bring the friars and
their teachings directly to any group of islanders. Mendaña's
colloquy with Bilebanarra, largely in sign language, was the only
recorded event of the kind. It has been suggested that the friars'
purpose was actually a preliminary survey of the Great South
Land with the intention of returning later in large numbers. To
establish an extensive Franciscan mission did, in fact, become
one of the planned objectives of the Spanish in the Pacific.[8]

In mid-February Mendaña despatched Sarmiento with
sixteen soldiers and six bearers on a four-day reconnaissance
into the mountains. Their attempt at an unobtrusive pre-dawn
departure was seen and the cries of large conch shells sounded
through the hills. Hundreds of armed natives appeared along the
trails. Unnerving as was the situation, the Spaniards' efforts at
negotiating a peaceful passage through the gardens and villages
seemed to work. Nevertheless, the explorers spent the night
on a ridge, firing two arquebuses every quarter of an hour to
make clear that they were awake and watchful, and by morning
Sarmiento had decided to return to the ships.

Bilebanarra appeared to lead them back to his territory, but there they found a large crowd of warriors waiting with arrows fitted to their bows. Thoroughly alarmed, Sarmiento tried to seize Bilebanarra as a hostage, but he escaped. A second man was seized, but on this the others let loose flights of arrows. The Spaniards responded first by firing into the air, but when this had no effect and a soldier was wounded, began shooting at their opponents. As one wounded man attempted to rise, Sarmiento killed him with his sword, whereupon the attackers fled. Enraged, Sarmiento burned houses and shrines, and with an uncle of Bilebanarra as prisoner, marched onto the beach, where the ships' boats were waiting, alerted by the sound of gunfire.

Sarmiento reported on the country he had seen. He noted an abundance of food—roots and fruits, pigs, 'fowls of Castile' and various birds. He spoke of 'indications of gold'—probably iron pyrites, which are plentiful in the sands of the streambeds.[9] There were pearls, but not in quantity. Mendaña, however, was very angry at Sarmiento's resort to violence and brutality.

Nevertheless, Bilebanarra's uncle was kept on board for three days. When no one appeared to seek his return, Mendaña sent Sarmiento with 30 soldiers to take him back to his people along with several objects Sarmiento had allowed his soldiers to pilfer. Terrified islanders met them, received the uncle with joy, and left a pile of coconuts and taro roots on the ground for the Spaniards.

The question of whether Santa Isabel was an island or a mainland remained. On 4 March, Mendaña despatched the campmaster, Pedro de Ortega, with a large armed party on an eight-day expedition up a rugged central ridge. Making their way through swamps and thick bush, they crossed the same river more than fifteen times before they began the ascent of a steep, stony mountainside. Repeatedly they were met with hundreds of armed men, sometimes assailed by shouts of 'Fuera! Fuera!' ('Away! Away!'), a word the natives had learned from the Spaniards. At other times they were shot at with arrows. Despite their wounds, the Spaniards marched stolidly on, for any appearance of fear brought on intensified attacks. They took one chief prisoner, and on questioning him Ortega was given to understand that Santa Isabel was an island. The man drew a circle on the ground. Within the circle was land, all around was sea. To confirm this, Ortega sent one Gaspar de Colmenates with eight soldiers to a very high ridge. From here the men saw the sea extending north and south, with only the view to the west obstructed. Their report satisfied Ortega and in a descent that was virtually a running battle with attacking islanders, and drenched by heavy, continuous rain, the group returned to the coast, where 'they all arrived very tired and wet, and covered with mud'.[10] One wounded soldier died.

Altogether seven reconnaissance patrols appear to have been sent from Bahía de la Estrella into the island's interior or along its shoreline. There could be little doubt that Santa Isabel was an island.

There is no record of women being violated, despite the details given of other excesses. While native women generally remained in the background of any encounter, there were occasional meetings. But between the discipline maintained by Ortega and the undoubted exhortations of the friars, neither soldiers nor sailors seem to have transgressed in this respect. Any such incidents, however, may have been omitted from the journals.

At the beginning of April, 54 days from the start of her construction, the brigantine *Santiago* was ready for sea. She was undecked, just sufficiently large to carry 30 men, their stores and ammunition, and a small culverin. Boarding nettings protected her from attack. On 7 April, under the command of Ortega and with a captive native as an interpreter, some twelve sailors and, by one account, eighteen soldiers embarked. As chief pilot, Gallego began directing the exploration of the two parallel chains of volcanic islands and coral reefs that constitute the Solomons. From time to time the Spaniards' hopes rose as an outline on the horizon or a feature of a coastline seemed to be evidence of a large mainland, only to be disappointed as they investigated further. Encounters with the indigenes varied from the amicable to the intensely hostile. More than once an initially friendly reception changed suddenly into a fierce attack when the Spaniards asked for food or as the brigantine departed, which was evidently seen as cowardice.

At one island several hundred warriors in some 20 canoes attempted to seize the brigantine, hurling arrows and stones. Arquebuses were fired in response, many attackers were killed and the others withdrew only to rally and attack again 'with more fury'. Beaten off once more, they retreated to a nearby ridge with loud shrieks and yells. Ortega's subsequent effort to make peace failed. The native fighters were people of great strength, Gallego noted, and fought furiously.[11] On another island the natives gave them a pig, and the brigantine's people celebrated Easter by eating 'the first fresh meat we had tasted since we left Peru'.[12] The Spaniards then seized three canoes, two of which were exchanged for two more pigs. The seizure of canoes to be returned for pigs served as an effective ploy a number of times.

Despite errors in his estimates of latitude and distance, Gallego's work among the islands was commendable. Santa Isabel was circumnavigated. Other principal islands and numerous smaller ones were recorded, among them Guadalcanal, apparently named for Ortega's birthplace in Valencia, as well as Malaita, San Cristóbal, Florida, Choiseul and the volcano-island of Savo, which was erupting. A number of the pilot's names remain on today's maps. On 5 May Gallego and Ortega returned to Bahía de la Estrella, and from the results of this survey the Spaniards had to conclude that they had found an archipelago, not a continent.

The situation on Santa Isabel showed some improvement. One of Bilebanarra's brothers and his men slept overnight on board the *Los Reyes*, while a Spanish lad spent two happy days with the natives. Both sides had learned enough of the other's language to make possible a degree of communication. However, promises were made and not kept. Food required by the Spanish continued to be a potential flashpoint. Undoubtedly the fragile economy of the island simply did not allow for suddenly provisioning over a hundred visitors who contributed nothing. The cost of this to the islanders was something the Spaniards probably did not fully understand. Gathering quantities of roots and coconuts was a major task in itself. Whether the pigs were fully domesticated is not clear; they may have had to be captured. In some parts of Melanesia there were, in fact, cults in which pigs were central. Meantime, the search for gold and spices had yielded only some wild ginger and basil, and a bark that tasted a little like cinnamon.

Having arrived at the beginning of the region's wet monsoonal season, the Europeans were also contending with extreme heat, high humidity and almost continual heavy rain. Many had come down with fever, most likely malaria, with four soldiers dying at Bahía de la Estrella. Fresh foods were certainly protecting them from scurvy, but the swarms of anopheline mosquitoes were another matter. It would be three centuries more before the mosquito-borne parasite was discovered and the connection made with human illness. Now, however, with Santa Isabel understood to be merely another island, it was necessary to continue their explorations.

Three days after the brigantine's return to the ships' anchorage, the expedition abandoned Santa Isabel. The islander who had served Ortega as an interpreter had escaped, and Mendaña took two others of Bilebanarra's men to serve in that capacity. He had dismissed a further two, embracing them, providing gifts and leaving word for the chief that as he had not visited the Spaniards for some time, he, Mendaña, was taking two men to show to the King of Castile and would bring them back. 'I said this,' wrote Mendaña, 'that he might not feel aggrieved, and regard my action as a breach of our peace and friendship.'[13] It was scarcely an act of honest friendship. However, the abduction of natives must also be seen from the Spanish point of view. They were needed as interpreters, but much more importantly, while they were being deprived of homes and families, they were being taken from barbarism to conversion and the salvation of their eternal souls. To the 16th century Spaniard this was full justification for their abduction.

Chapter Four

GUADALCANAL

15 89

The expedition's objective now was the rugged, rainforested mountain mass of Guadalcanal, which had previously been seen from the brigantine. Catoira noted that it exceeded all their other landfalls in size and beauty.[1]

The journey from Santa Isabel took five days, with alternating calms and strong contrary winds. By 14 May, Gallego had found a sheltered harbour, which can with certainty be identified as the site of the Solomon Islands' modern capital, Honiara. Mendaña named it Puerto de la Cruz. Several canoes came out to the ships, while over 1000 people crowded the beach. Unafraid and apparently friendly, their chiefs boarded the *capitana*, where they were hospitably received and given gifts of beads.

Mendaña and the friars disembarked. A cross was erected on a hilltop within sight of the ships, mass was said and 'Vexilla

Regis Prodeunt' sung. To the islanders who came to watch, Mendaña made signs of peace, embracing them and attempting to explain that he would do them no harm. As the Spaniards turned to leave, however, the onlookers began shouting, flourishing their weapons and firing off arrows, while an additional 300 or so were seen approaching through a valley. Mendaña ordered gunfire. Two warriors were killed and the rest scattered and ran. By evening the Spaniards saw from the ships that the cross was being carried away, although the next day Sarmiento and a landing party found that it had been returned to the hilltop. They re-erected it and Mendaña took possession of the island in the name of King Philip II.

The next few days were quiet. The brigantine was leaking badly and, well guarded, was beached and caulked. Catoira mentions a church, probably a thatch structure. Mendaña, with his officers and fifteen soldiers, headed inland. Their exploration was brief, but they saw with astonishment the extent to which the island was populated. Villages were scattered over the hills. There were groves of palms, plantains and other trees, and cultivated fields irrigated by little streams issuing from the forests. Spanish records citing large populations in the Solomon Islands have been questioned, but the extent of cultivation and the virtually countless villages chronicled by the explorers make their claims entirely plausible.

On 19 May a party of 22 soldiers and their servants, under Andrés Nuñez, set off for the interior on a six-day exploration trip. In the interval Mendaña himself made another sortie into the hinterland. He noted the 'marvellous' number of villages on hilltops and 'some beautiful plains'. Scarcely any people were seen and Mendaña's group returned to the ships without, as

Catoira noted, 'doing any harm'. On the beach, however, there had been a brief skirmish, the Spaniards capturing a boy about six years of age and taking him to the ship.

Food was becoming a serious issue. Rations to the Spaniards now consisted of eight ounces of biscuit and half a pound of salt meat, some of which was beginning to rot; 'our roots were finished and our people were worn out, both from having little to eat and from sickness'.[2]

Every effort to barter for food had failed. Calling out to people and displaying trade goods brought no response. The huts in the villages were deserted. Sarmiento, foraging, found only small caches of food, while the four men waiting with a boat for his return to the beach were attacked by some hundred warriors. Firing at them and killing one drove them off, but not before one soldier had his leg felt—as he was convinced—for tenderness, by a fighter bent on eating him. Such provisions as had been found were ferried to the *almiranta*. Shortly afterwards Andrés Nuñez and his party arrived, exhausted but carrying edibles which were taken out to the *capitana*. Their tale was one of repeated attacks, ambuscades and occasional concordant trading of trinkets for food. Their efforts at prospecting failed. Swift-running rivers made panning almost impossible, and the miners were easy targets from the higher banks or hillsides.

Violence escalated. A watering party guarded by two arquebusiers was attacked as they were loading water jars into their boat. Totally overwhelmed, the entire group was slain and their bodies dismembered, some beheaded, others with tongues sliced off and skulls opened for the brains within. Only Gallego's black servant escaped, diving into the sea and as he swam defending himself with a cutlass from those who pursued him into the water.

The Spanish revenge was fierce. Sarmiento burned hundreds of houses and his men shot, killed and wounded where they could. After a later clash Mendaña hanged, beheaded and quartered the natives who were killed or captured, leaving their remains where the watering party had been slain. Under the burden of continued hostility, hunger, infirmity and death, Mendaña's earlier humanity seems to have crumbled. There were evidently no further consultations with the friars on the fairest way to carry on relations.

Yet the small number of Spaniards with their few arquebuses, heavy, awkward and slow to fire, could never create more than a minor impact upon the hordes that descended upon them and vanished into the mountains when pursued. The guns inspired fear, although in time the islanders realised that an arquebus fired into the air, which was usually the Spaniards' first move against them, did no injury. They employed camouflage, covering themselves with leaves and branches so as to be almost invisible among the bushes. They also learned that in attacking the brigantine from the water, they could avoid becoming easy targets by ducking their heads as they swam. In one attempt to get back a canoe seized by the Spanish to be exchanged for a pig, the islanders fashioned a dummy pig from straw and laid it on the beach. The 'pig', however, did not move, and the Spaniards held on to the canoe. There was treachery and trickery on both sides. Probably the greatest damage done to the islanders was not so much the killing, for they pursued their own wars, but the violent disruption of the normal pattern of their lives.

The day Andrés Nuñez and his party headed inland, the brigantine was again under sail, following the coast of Guadalcanal and investigating other islands. Aboard the brigantine there were some sharp disagreements between Henríquez and Gallego. Henríquez refused to allow his men to fire upon the natives until they had released their arrows, a chivalrous gesture that encouraged their assailants and infuriated the tough, realistic Gallego. At times, however, Henríquez' efforts at fair and peaceful trading brought welcome results—fresh water, plantains and yams. In Catoira's words, he 'gave them barter for everything',[3] and there were small rewards. On one occasion he was amazed by the voluntary return of a knife, apparently lost by some sailor. On another a captive assisted with loading a shrieking pig from a canoe onto the brigantine, whereupon Henríquez freed him and put on him a cap and several strings of beads. The two men embraced, almost in tears, and the native, returned to the beach, stood there and signed his thanks. Henríquez attempted repeatedly to obtain food peaceably, but failing, resorted at times to having his men scare off the inhabitants with gunfire. He then seized any available food, leaving strings of beads in payment. At other times the encounters degenerated into battle. Catoira commended Henríquez' firm leadership: 'Don Hernando ruled well in all things in war as well as in peace'.[4]

On the island of Malaita great excitement gripped the Spanish soldiers when they thought they had found gold. The warriors of the island apparently fought with clubs with round stone heads wrapped in woven grass. The weight of the stone convinced the soldiers that here indeed were weapons of gold, and trading in caps for clubs was lively until Henríquez ordered it stopped. Caps were valuable trade items, to be exchanged only for food. When the men continued their bartering clandestinely, Henríquez smashed together two stone heads until they broke. No one knew what kind of stone this was, but it was not gold.

The notion of gold clubs, however, did not go away. In the inns and taverns of Callao the rumour lingered for years.

The brigantine returned to the anchorage on 6 June. No one was missing or wounded, but five or six men were incapacitated by fever, one of them Gallego. They had, nevertheless, discovered another six islands. Now preparations were begun to leave Guadalcanal. Thirty-eight men were ill and one had died, but the voyage had to continue.

If Hernando Henríquez had managed to retain some of his sense of fairness against an enemy, Mendaña seems by now to have lost his completely. On 12 June he made a final vengeful foray along the shore and inland, burning villages. On Sunday 13 June the entire ships' companies landed to hear mass, and at midnight the expedition sailed. They left their cross and their church still standing, and a population that was without doubt vastly relieved to see them go.

For a week the vessels tacked repeatedly in rough seas, pursuing an east-southeasterly course against unknown currents and southeast trade winds. One of the pilots died and was buried at sea. The brigantine, under tow, reared and plunged dangerously. The weather worsened. In gale-force winds a mizzenmast yard broke. A sailor holding on to the mast as he tried to furl the sail lost his grip and fell from the top into the sea. Catoira wrote: 'when we rushed to look for him we found him clinging to some ropes that were dragging through the water, and he was so handy and agile that he climbed up without assistance'.[5] Juárez Méndez had survived his second plunge into the sea.

On 30 June they entered a protected harbour at an island they named San Cristóbal. Their reception near a village of some 80 houses was peaceable, and Mendaña took possession

in the name of Philip II. The next morning the Spaniards came
ashore in two groups, determined to forage. Gallego recorded
the events: 'the Indians . . . went to arms, making signs to us
to re-embark . . . I saw one of the chiefs making exorcisms
and incantations to the Devil . . . it appeared that his body was
possessed by the Devil'.[6] Then with loud cries, brandishing bows
and arrows, darts and clubs, the warriors rushed forward. 'We
were forced to fire on them, and we killed some and wounded
others, whereupon they . . . fled.'[7]

Chapter Five
THE LAST ANCHORAGE

SPE ET METV.

15 89

With the two ships at anchor, Mendaña sent a presumably recouperating Gallego with the brigantine to explore and to gather provisions. The men coasted to the far southern end of San Cristóbal. In an amicable encounter with a group of natives, the Spaniards were told that there was land to the south-east, possibly the Santa Cruz Islands. But a sailor up a tall palm tree saw only an empty horizon, while heavy swells from the east suggested that there was little if any land in that direction.

There was sometimes a curious mixture of regard and savagery in the conflicts that took place. When a man and a boy came on board the brigantine, the Spaniards seized the man, but the boy leaped overboard and, clutching his bow and arrows in one hand, swam for shore. When two sailors followed, he threw his weapons in their faces, winning the admiration of the seamen for his courage. The lad was taken, but subsequently he and the man were traded for pigs. Each was embraced by the Spanish commander, given beads and sent off 'well pleased'. There was contempt and bravado as well. Five boys the sailors were attempting to take into the brigantine wrestled themselves free and escaped to shore. 'On reaching it they began to make threatening gestures . . . jeering at them [the sailors] . . . and turning their hinder parts towards them with unseemly impudence.'[1]

The Spaniards showed their own bravery. Their arquebuses failing them in one fight, they charged with swords and hatchets, their captain, although pierced through the arm with a dart that stood out on the other side, shouting, 'Santiago! And charge!' There was courage too in the marches made into strange country by groups of 20 to 30 men, knowing they could be attacked by crowds of warriors many times their number.

In the early evening of 14 July the brigantine returned to the ships with the injured from their exploratory forays. The following day the captains, pilots and officers met on board the *capitana*, to hear Mendaña's orders to prepare as speedily as possible for further exploration. But after the months at sea, the ships' hulls had first to be cleaned.

The several accounts of the voyage do not agree entirely on the details of events that followed, but it is clear that the cleaning and repair of the ships, 'with the little pitch and rigging we had left', was the principal undertaking.[2] At daybreak on Saturday 17 July, Mendaña landed with the men from the *capitana*, 'and all the chests and stuffs . . . were brought on shore and placed in a large shed', probably a native canoe shelter. 'And all the men from the *Almiranta*, with the Master of the Camp and the stuffs, occupied another shed.'[3]

With the men quartered in some of the deserted native huts, the huts nearby were burned to create open ground around the camp. The men were strictly forbidden to venture out of the area, for they could see that they were being watched from the heights. No doubt the islanders now assumed that the intruders were about to settle permanently, and at dusk that evening a warlike crowd was seen descending from the hills. The artillery—the ships' culverins now on shore—was being set up when a large canoe filled with well-armed warriors was sighted near the harbour. A gun was turned upon them and fired. The ball hit the water alongside the canoe with a huge splash, in one account, first flying over the heads of the men and then plunging into the water. With its occupants paddling frantically the canoe headed straight out to sea. The artillery was then discharged towards the mountains where the warriors were descending, with reports that echoed among the hills. The attacking force broke up and fled and, heavily guarded, the camp was left undisturbed.

Two days later, no doubt at high tide—and perhaps on two or three successive tides—the empty ships would have been run onto the beach. Long tackles attached to the tops and looped around large wooden stakes driven into the ground were hauled tight, and the vessels heeled over. The crews went to work, for the next three weeks scraping hulls thickly encrusted with barnacles and seaweed, heating pitch in iron pots over beach fires, caulking, paying seams with the pitch, and making the most urgent repairs on the rigging and elsewhere.

Despite the security won by the culverins, some of the soldiers invited trouble, unable to resist slipping out from the camp to shoot pigeons or seek other fresh food. During early mass one morning two men left to cut palmetto and were ambushed. One, although hit by a dart, reached the camp, his cries raising the alarm. The other, who was relieving himself, had tripped on his falling hose. He was found 'pierced through and through', his skull split by the blow of a club. Mendaña threatened his men with hanging or stabbing if they again ventured out of the camp. To reinforce this order the arquebusiers were deprived of match for their guns except when skirmishes seemed imminent.

As work on the ships was completed, Mendaña seems to have conferred with his captains and pilots on whether to settle on the island or to continue their explorations. He then ordered a *parlamento* at which each member of the expedition would be free to express his opinion.

On Saturday 7 August, 58 men—captains, pilots, soldiers and sailors—gathered. Mendaña spoke first, saying he wanted to sail some 150 leagues farther, to 20° or even 30° South, where it was understood there was land. Should none be found, they could still serve God and the king by returning to the best of the islands they had seen and establishing a colony. Gallego spoke on the condition of the ships. Depite repairs, they still leaked and were badly seaworn, while seeking provisions with the brigantine was no longer feasible. Their search for food was known throughout the islands, and everywhere supplies were hidden from them. Sailing south with little food and their few remaining containers of water, they would be lost. On settlement he offered no opinion. Others did, however. Settlement was part of Mendaña's instructions, but it was clear to everyone that the natives were in overwhelming numbers and the explorers few, with many of them now sick or wounded. Lead and match for the arquebuses, on which they depended for survival, were almost gone. Locks were broken and guns otherwise beyond repair. Attempting to establish a colony under these circumstances, said the friars and the campmaster, Ortega, would be a disservice rather than a service to His Majesty.

The miners, however, were certain that there was gold to be found, and voted for settlement, some of the soldiers agreeing with them. Sarmiento also urged forming a colony. If they did return to Peru, however, they should pursue a southeasterly course to find the land that he had originally proposed, islands somewhere west of Chile. Mendaña favoured this; the lure of a rich new land remained strong. By waiting a month for the September equinox, they would certainly have a favourable change of wind. In the meantime the brigantine could go in search of more provisions. The pilots, however, 'would approve of none of these things, saying that the landsman reasons and the seaman navigates'.[4] They wanted an immediate departure, and pressed for a northerly course that would take them to New Spain. A northern transpacific route from west to east had recently been pioneered by the Spanish, and the pilots would have been aware of this. A compromise appears to have been reached. They would depart promptly and sail to the southeast as the wind permitted.

Mendaña wrote: 'As we had with us only two Indian boys as interpreters, it seemed to me that this was not enough . . . if these should die'.[5] Thus in a night raid a party of soldiers seized a man, his wife and child, and the woman's sister. Mendaña objected to the women, but evidently the family asked to remain together. Mendaña ordered them clothed. They were allowed on deck during the day and for the women's protection locked below at night, the key given to the vicar. Apparently there were no untoward incidents. In time the family was said to show 'no sign of sadness, and laughed with everyone'.[6] The Spaniards now had six potential interpreters. That these captives would speak the same language as any other people they encountered was questionable.

Chapter Six

THE JOURNEY BACK

I n the quiet pre-dawn darkness of 11 August 1568, anchors were raised, the sails filled and the little fleet quitted the island of San Cristóbal after a stay of 41 days. Against heavy winds they beat southwards, covering only 30 leagues in seven days. The brigantine had to be sacrificed. Under tow in rough seas, it was nearly smashed against the *capitana*. Regretfully, Gallego cut it loose. On the 17th they doubled San Cristóbal's southern cape. In fierce winds and heavy rain the *capitana*'s main yard broke and the sail was shredded, while the *almiranta*, now with Sarmiento on board, was briefly lost from sight. A clash of wills erupted between Gallego and Mendaña, the chief pilot maintaining that to pursue a course to the southeast was madness. It was imperative to cross the Equator and follow a northerly route to the Americas. Shouting across the water, the pilots on the *almiranta* were consulted and their agreement received. A formal document was drawn up on the *capitana*, and a similar statement, tied to a line, was flung across from the *almiranta*. Mendaña yielded. He warned his pilots, however, that they would suffer from this decision.

A course was laid for the North Pacific, California and New Spain. On 6 September they crossed the Equator and eleven days later sighted an atoll in the Ralik chain of the Marshall Islands. Ortega and Henriquez landed to find a deserted village, the people apparently having fled out to sea. They found no fresh water and the only object of interest was a piece of iron, probably a small nail, tied to a stick. A European ship must have been in the region at some previous time. On 2 October at 19° 18' North, 166° 38' East, they came upon the bush-grown islets and barrier reef that constituted a solitary island they named San Francisco, now Wake Island. They sailed all around it looking for signs of fresh water, and seeing none, resumed their northerly heading.

Sightings of the *almiranta* became sporadic and after 16 October the vessel was seen no more, despite the *capitana* having waited under reduced sail and even heaving-to for some 24 hours. Mendaña concluded angrily that Sarmiento had prevailed upon the captain, Ortega, to alter the *almiranta's* course.

That afternoon a violent storm burst upon the flagship. Heeling to port, the *Los Reyes* was half under water, her people tumbling against the bulkheads amid screams and prayers, some of them forced to swim inside the ship. Somewhere in the chaotic darkness, one of the friars bravely led them in reciting the Creed, urging them to die like Christians. Above them the stern cabin was carried away. The helmsmen clung helplessly to a demented whipstaff. In 45 years at sea Gallego had never seen such a storm.

Finally the men chopped down the mainmast, which slid overboard with the tangled mass of maintop, yards, rigging and sails, and the ship's boat, flooded, was shoved into the sea. Slowly the ship righted itself a little. At the bow Catoira and some helpers rigged a blanket as a kind of sail while a bonnet was shaped into a small stormsail. Inside the ship one of the pilots fell into the water. Catoira wrote: 'he began to swim, but, being heavy, he was going to the bottom and would certainly have drowned, had it not been for Cordovylla, a negro, who threw himself in and got him out'.[1] The storm abated, but three days later the ship was again engulfed in terrible weather, 'as if the world was coming to an end'. Yet, as the men believed, God miraculously delivered them once again.

Conditions on board were becoming desperate. For many days the men's rations had been eight ounces of mouldy biscuit, half a pint of water swimming with cockroaches, and 'a very few black beans and oil'.[2] A rain shower sent them rushing to catch enough water for three days, probably by creating a large basin with a sail. Almost daily someone died and was thrown overboard.

Scurvy set in, as did a curious blindness. Terrified soldiers clamoured to turn back, but with an unusual show of authority Mendaña was able to calm them, explaining the insanity of doing this, promising them that the American coast was not far off. Then, alarmingly, the wind died. Just at this point a large piece of timber was seen drifting with the current. A sailor jumped overboard and retrieved it—a clean, fresh, pleasantly scented length of wood. Eight days later land was sighted, to the nearly hysterical joy of the ship's company. At about 30° North they saw modern Mexico's Baja California. Having jettisoned the ship's boat in the storm, they now put together a raft in order to land on a deserted shore and bring on board casks of fresh water. On 23 January 1569, the *Los Reyes* entered the harbour of Santiago de Colima on the Mexican coast. Two days later the *almiranta* came in, also dismasted. On 4 April the *capitana* reached Realejo, now Corinto in Nicaragua, and by pledging all his—and apparently some of his officers'—private property and borrowing 1400 pesos from Gallego, Mendaña was able to have the ship repaired and victualled before resuming the journey south.

The enmity between Mendaña and Sarmiento flared in the weeks and months that followed. At Colima and Realejo, Sarmiento brought official indictments against Mendaña for his handling of the expedition. Each time Mendaña had Sarmiento arrested and each time Sarmiento secured his freedom. He did not return to Peru with the ships, but travelled separately to Lima where he began a campaign of letters to the king and evidently interviews with the viceroy to denigrate Mendaña and promote himself as leader of a new expedition to the Pacific.

On 11 September 1569 Mendaña's two sea-battered ships arrived at Callao. The voyage had taken 22 months. One account records that '*al pie de cuarenta hombres*', almost 40 men, had died, a loss of between nearly a half to about a third of the expedition's complement.[3] Names and the cause of death of most were listed by Catoira:

> Alonso Martín, soldier, died on the first island of Santa Ysabel of convulsions from a wound dealt him by the *yndios* . . . Diego

de Frias, of fever on the said island . . . two slaves of the chief
pilot . . . Benito de Aguilar, old soldier from Peru . . . Diego
Sanchez, sailor of the *capitana*, on the return . . .[4]

And on it went.

The family taken from San Cristóbal was landed safely.
Those of them who died later in Peru did so, as Mendaña wrote,
as 'devout Christians, invoking the name of Jesus many times.
Many thanks should be rendered to Our Lord'.[5]

What had the expedition achieved? It had recorded the
existence of the Solomon Islands, returning with considerable
detailed information on a hitherto unknown Pacific archipelago. The Ontong Java atoll or the Roncador Reef, probably
Nui in modern Tuvalu, and Wake Island had been tentatively
added to contemporary knowledge of the Pacific, as had a
somewhat better understanding of the ocean itself. What of
Ophir? No one on the voyage could have imagined that they
had found it. Mendaña, writing to Governor de Castro, referred
to 'the discovery of the islands', implying that they had located
the islands they sought, but made no claim that these were the
golden isles of tradition.[6] Sarmiento was vehement in his denial
that the islands, 'vulgarly but incorrectly called the Solomon
Islands', were related to Ophir,[7] but the name was soon routinely
employed, even by Sarmiento himself. What of the existence of
gold? Gallego, Mendaña and others maintained that natives had
both reported and recognised it. With little doubt, the stories
told by the returned adventurers were gilt edged. Perhaps they
had not discovered Ophir, but there were traces of gold that
could have been found had circumstances been different. And
Ophir itself could not have been very far away.

With the exception of a reference by Gallego to a 'chart which I shall add' to his written account of the voyage,[8] the narratives of Mendaña's expedition do not mention the preparation of maps. Charting would have been Gallego's responsibility, and in the normal course of events his maps would have been sent to Seville to the Casa de Contratación de Indias, literally the House of Trade in the Indies. The Casa de Contratación had extensive powers over maritime affairs as well as commerce, including recording new hydrographic information on an official master chart. Evidently Gallego's chart no longer exists, but certain later maps were probably copies. At the end of the 1500s there were a considerable number of charts on which the Solomon Islands were recognisably drawn and reasonably accurate in placement. But to the south, sweeping across the entire width of the Pacific Ocean lay the huge continental expanse of Terra Australis Incognita. The vast, mysterious South Land remained undiscovered.

Chapter Seven

THE SECOND VOYAGE:
1595–1596

I n terms of the outcomes which were expected from such a voyage, Mendaña had accomplished little. He had created no colony, found no spices or precious metals, made no converts to Christianity. He had discovered no vast southern continent to yield wealth and power to Spain and to himself as a *conquistador*. Yet Mendaña's belief in Terra Australis Incognita was unshaken. Perhaps driven by the very fact of his failure, he became obsessively determined to return to what he believed to be islands edging the coast of the continent and there to plant a Spanish colony. From this base the discovery, colonisation and Christianising of the South Land would take place. In command of this enterprise, potentially so rich in souls and wealth, Mendaña saw himself. This objective became the single-minded centre of his life.

Thus in his report of the voyage addressed to King Philip on 11 September 1569, Mendaña described the Solomon Islands as attractively as possible—the large populations, thousands of souls in need of salvation, and the desirable products of the islands: 'cloves, ginger, and nutmeg . . . They also say that there are pearls . . . also that there is gold'.[1]

However, Mendaña now faced the first of the seemingly endless obstacles that would frustrate his plans for almost 25 years. His uncle, the Licenciate Lope García de Castro, had returned to Spain from Lima and Mendaña was confronted by the new viceroy, the distinguished soldier and diplomat Don Francisco de Toledo.

Toledo's twelve years in office would become among the most significant in Spanish colonial government. A man of energy and tenacity, he carried instructions from Philip II that enabled him to introduce meaningful improvements in administration, to establish order in the silver mining industry and the Crown's taxation procedures and to curtail the independence of the church in the Americas. In 1571 he crushed the final Inca bid for a separatist state with the execution of the last Inca emperor. In effect, he consolidated the position of the Crown of Castile as the sole authority in the colony. The faraway settlement proposed by Mendaña had no place in Toledo's tightly woven scheme for concentrated regal power.

Toledo made it clear that he would consent to no expedition without the express approval of the king, and in a letter to the monarch he dismissed as costly and wasteful attempts to Christianise natives in distant regions, when thousands of indigenous people in nearer locations awaited conversion. Toledo would have been aware of the degree to which Spain's lines of communication and trade were already over-extended, and of the drain in men and money imposed by the administration and defence of a sprawling empire.

The exact sequence of events at this time is unclear, but it soon became obvious to both Sarmiento and Mendaña that there was no possibility of another voyage into the Pacific. Sarmiento, however, had gained some favour with the viceroy, and in 1570 became part of Toledo's entourage on a wide-ranging four-year tour of the viceroyalty, which included the final destruction of the Inca dynasty. Mendaña, on the other hand, had somehow incurred the viceroy's personal hostility. In a later report to the king, Toledo would scathingly denounce both Mendaña and Castro as liars and charlatans.

Realising that his proposals would win no support in Peru, Mendaña left for Spain to approach King Philip directly. On the six-month journey he carried 500 signatures of men prepared to join in his conquests. He probably arrived in Madrid in August 1571.

For two years Mendaña hovered about the royal court in Madrid, repeatedly submitting in the form of letters to the king petitions to the Council of the Indies for a *capitulación*, in effect a contract, allowing and providing for a settlement in the 'Islands of the West', that is, the Solomon Islands. He requested four ships costing less than 60 000 pesos, and would himself provide 500 men and derive his own remuneration from the established colony. Should negotiations fail, he asked only for suitable recompense for his past services. Apparently there was no reply. His uncle, the Licenciate Castro, was in distant Bolivia on royal business and unable to exert useful influence. Mendaña simply continued submitting his petitions, varying his requests

from time to time. At one low point he abandoned the idea of
a colony and asked only for a gratuity that would maintain him
as befitted his rank as a general. Another time he stressed mainly
the spiritual welfare of the islanders.

The Council was receiving letters from other sources as well:
from Toledo, from Dr Gabriel de Loarte, President of Panama
and an enemy of Castro, and from Sarmiento, all harshly critical
of Mendaña. Nevertheless, Mendaña's proposals finally received
royal sanction, and the slow process of being granted the necessary documentation ground on, eventually with the support,
financial and otherwise, of Castro, returned from Bolivia and a
member of the Council of the Indies. On 27 April 1574, Philip
II signed a *capitulación* with Mendaña 'for a voyage of discovery,
settlement and pacification of the Western Islands in the South
Sea'.[2] The expedition was to establish a colony.

The document was comprehensive. It provided for the necessary ships and crews and provisions for one year. It specified 500
men, 50 of them married with families, and cattle, horses, goats,
sheep and pigs. Three fortified cities were to be founded within
six years, Mendaña to provide 10 000 ducats as security that
these projects would be carried out. In return he was granted
sole jurisdiction in the colony for two generations, with slaves,
exemption from customs duties for ten years, power to bestow
repartimientos, that is, the labour of natives, to his followers and
the authority to coin gold and silver.[3] He would be answer-
able only to the Council of the Indies, and a later satisfactory
inspection and report would provide him with 'a vasallage in
perpetuity and the title of marquis'.[4] Two additional documents
clarified the first, and a special decree gave him the power to
nominate anyone he chose as his successor.

Yet it was two more years before bureaucratic procedures were complete and Mendaña was ready to sail for Panama. Aware of the enmity he would face from Loarte in Panama and Toledo in Peru, Mendaña repeatedly requested from the king a document instructing these authorities not to impede his journey. A missive from the king to the viceroy required that Mendaña receive 'board befitting his rank', but it is evident that this had little effect on either Loarte or Toledo. Mendaña also made a very different request: he asked for the return to him of a native he had brought from the Solomons, evidently to take him back to his island. Neither the identity of the man nor the result of the request is known.

At the beginning of 1575 Mendaña left the court in Madrid. He sailed from Seville apparently after August 1576, probably with the regular transatlantic convoy, and arrived in Panama sometime before mid-November.

Unbelievably, it was another nineteen years before Mendaña was able to embark on his expedition. Although a Crown-designated *adelantado*, that is, an expedition commander and governor of distant lands, he faced from Loarte in Panama and subsequently from Toledo in Lima unrelieved opposition and humiliation, even brief imprisonments in both places.

Then a sudden crisis put his ambitions in further abeyance. In September 1578 a little ship of about 100 tonnes emerged from a stormy traverse of the Strait of Magellan and, sailing north along the South American coast, began an unprecedented series of raids and depredation on stunned, unsuspecting and utterly unprepared Spanish ships and communities. Among the targets was the port of Callao, where the attackers sailed among the anchored ships, searched them for treasure and cut

their cables to set them adrift. By the time Toledo, riding from
Lima at the head of 200 men, arrived, the raider was out at sea.
Two years later the *Golden Hind* under Francis Drake returned
to England laden with a spectacular fortune in treasure taken
from the Spanish in a part of the world they had believed was
safely and indisputably their own. Toledo rallied every available
means to counter further attacks by Drake. Mendaña joined in
planning coastal defences. Sarmiento was despatched with two
ships to guard the Strait of Magellan. Captured by the English,
Sarmiento was taken to England where, set at liberty, he reportedly spoke in Latin to Sir Walter Raleigh and Queen Elizabeth
on his plans for a Pacific island settlement. Ransomed by the
Spanish Crown in 1590, he died in Lisbon two years later.

In December 1580 Toledo's term of office ended and he left
in May the following year. In Panama Dr Loarte had died. Freed
of his enemies, Mendaña pushed on with arrangements for his
voyage. Initially Drake's attack had repercussions on Mendaña's
project. It was thought that no South Sea settlements should be
established since, if seized by the English or other intruders, they
would offer 'succour'. However, four years later opinions had
changed. In a letter to the king dated 21 March 1582, the president of the *audiencia* of Charcas, Bolivia pointed out that with
ships guarding the Strait of Magellan, invaders were likely to
enter the Pacific by way of India. A Solomon Islands settlement
could raise the alarm for Peru. The following year the Charcas
authorities warned Lima of the possibility of the English occupying the Solomon Islands.

In January 1587 the English returned to South America.
A squadron under Thomas Cavendish bombarded and looted
Peruvian coastal towns and captured the galleon *Santa Ana*,

with her 122 000 pesos' worth of gold. In an atmosphere of heightened concern for Spain's claims to the unexplored South Pacific, Mendaña pressed on with his preparations. Documents record the enlistment of volunteers and the raising of funds through contracts that promised contributors such key positions as captain and sergeant. Mendaña was actually in serious financial trouble. On his return from the voyage to the Solomons, in 1569 he had gone into debt in Nicaragua to pay for desperately needed ship repairs. Life in Madrid had also been costly. Numerous letters to the king contain pleas for payment for expenses incurred at Corinto, for an income, grants, loans or remission of moneys he owed to the Crown.

In May 1586 Mendaña had married. Isabel Barreto (or Bareto) was the daughter of Rodríguez Nuño Barreto, listed as among the first *conquistadores* and *pobladores* or founders of Peru, and his wife Doña Mariana de Castro, who were apparently people of some social standing. No specific description of Isabel appears to exist, but she was evidently one of at least nine children, close to her brothers and sisters and, as circumstances would reveal, a young woman of great determination, ruthlessly ambitious and absolutely self-serving.

Mendaña would then have been in his mid-forties, deeply involved with his expedition, which was now receiving the favourable interest of a new administration. During the same month in which he was married he had useful discussions with the new viceroy, Don Martín Enríquez de Almanza, Count of Villardompardo. Despite the fact that he was in financial straits, with the ranks of general, *adelantado*, governor and captain-general of the land he conquered, and with the prospect of a future marquisate, all titles he could pass on to his heirs, Mendaña's future appeared impressive. No surviving letter or comment casts light upon the personal relationship between Mendaña and Isabel, but from what is known of her it would very likely have been his prospects that had attracted her interest, and probably that of her family. For Mendaña the marriage provided immediate financial assistance. In his discussions with the viceroy on 25 May 1586—presumably shortly after the wedding—Mendaña proposed spending his wife's dowry on the venture. The expedition's second ship, the *Santa Isabel*, was subsequently bought with money from the dowry.

In 1588, García Hurtado de Mendoza, Marquis of Cañete, took office as viceroy. He was impressed with Mendaña's venture, and from this point Mendaña received considerable support and cooperation from the viceregal government. In 1590 Mendoza appointed Mendaña *visitador* or inspector of the royal fleet.

The expedition was coming together. When the viceroy decided to reduce the cost of maintaining the government fleet by offering a number of vessels for sale, the galleon *San Gerónimo* went to auction and a month later was sold to Mendaña for 8000 *pesos corrientes*.[5] A condition of sale was that the ship be used for the colonising voyage, and it became the flagship, the *capitana*, of the expedition. In addition Mendaña obtained a frigate, the *Santa Catalina*, a small, fast sailer, and the galeot *San Felipe*, a galley-type vessel with sails and sixteen to twenty oars to a side. The viceroy provided guns and munitions: two small cannon, six culverins, 50 arquebuses, match-cord and 60 jars of gunpowder. The viceroy's own purpose was served as well. Once again an expedition provided the opportunity to rid the community of vagrants and other undesirables.

The fleet sailed from Callao on 9 April 1595. The names

of 354 men, women and children, settlers, sailors, soldiers and servants, are given, with a later document quoting a total of 378 persons. These were divided among the four vessels. Twenty-three children are listed for the *San Gerónimo* and six *muchachos* or youths for the *Santa Isabel*, now designated the secondary ship, the *almiranta*, of the expedition.

Heading north along the coast, the four vessels stopped at several small ports to pick up additional personnel and supplies, from the evidence arbitrarily requisitioning 'what was wanted'. There was little organisation. 'In each port there was disorder', Quirós, the chief pilot, commented later. In the harbour at Cherrepe the officers of the *Santa Isabel* noticed a new, strongly built merchant ship lying at anchor. In a private deal with the other ship's officers, men of the *Santa Isabel* arranged an exchange of vessels. Mendaña objected. But 'in order to gain their end, they secretly bored seven gimlet-holes in the ship [*Santa Isabel*]' and as she began making 'a great deal of water', the unsuspecting soldiers panicked, refusing to sail in a ship 'so unseaworthy'.[6] Mendaña gave in, mortgaging the ships to pay the difference in the value of the two vessels. He 'felt and complained much of this proceeding', Quirós noted, and 'threatened those who he believed to be the cause of it'.[7] But he took no known action. At Cherrepe, Lope de Vega, captain of the replacement *Santa Isabel*, enlisted a number of married couples, and at the nearby town of Santiago de Miraflores was himself married to Isabel Barreto's sister, Mariana de Castro. Mendaña gave Vega the rank of admiral.

The final stop on the coast was at Paita, where 1800 jars of water were taken on. Here Mendaña dismissed from the expedition several people he believed were not sufficiently respectable, and finished his recruiting with the hiring of a sergeant-major, who paid 2000 dollars for the post.[8]

On 16 June the squadron left the Peruvian coast, shaping a course to the southwest. Hopes and aspirations sailed with it, as did incompetence and jealousy, greed and incipient violence. With Lima's reprobates, a flawed choice of officers, a domineering wife and her family all seeking the mirage of an unknown Terra Australis the ships were, in fact, sailing towards death and disaster.

There were some stark differences between Mendaña's new expedition and his earlier voyage into the Pacific. The current expedition was a larger and costlier enterprise. It was graced with royal approval and was the outcome of a quarter of a century of dedicated effort on its leader's part. What, then, accounted for a journey that would be steeped in avarice, moral insensibility and random bloodshed in comparison with an earlier enterprise in which there was indeed violence, but violence mainly linked to survival? Mendaña, the novice explorer of 1567, wavering at times in his leadership, was now a man in his early fifties who had finally achieved the goal of his life, the opportunity to find the legendary antipodean continent of Terra Australis. Mendaña's dedication to this goal is beyond dispute. He must have felt a strong sense of triumph as he now viewed his squadron of four vessels under sail. But the question arises as to what further ability as a leader of men had he acquired in the intervening years, years spent at the court in Madrid, in Lima, in writing petitions and seeking the favour of great men. The young commander of 1567 had with him several men of real competence and, at least as they were conceived at the time, humanity and honour. He was not so fortunate on this second voyage.

The men Mendaña chose as officers for this expedition were widely divergent personalities. On 7 March 1595 he appointed as chief pilot of the squadron and captain and master of the *capitana*, the *San Gerónimo*, a Portuguese navigator, Pedro Fernández de Quirós. In the document appointing him, Quirós was cited as 'a person of great worth and trust and has wide experience and knowledge of seafaring'.[9] His combination of positions on board gave him authority ranging from being the final word on navigating the fleet to the overall handling and daily running of the flagship. He was probably in his early thirties. His narrative, edited and no doubt written in part by his secretary, the poet Luis Belmonte Bermúdez, is one of the chief chronicles of the expedition. It is evident that Quirós was indeed trustworthy and, within the limits of 16th century navigational expertise, knowledgeable. He was also someone who at least at times attempted pacific solutions to the confrontations that arose.

Two priests boarded the *San Gerónimo*: the vicar, Father Juan Rodríguez de Espinosa, and Mendaña's chaplain, Father Antonio de Serpa. The *almiranta*, the new *Santa Isabel*, also carried a priest whose name, however, is not recorded.

Several members of Isabel Barreto's family were on board the *San Gerónimo*. Her brother Lorenzo was 'captain of the infantry' and accompanied by two other brothers, Diego and Luis. With Isabel was their sister Mariana. They were a strong family unit and, through Isabel's position as the *adelantado*'s wife, privileged and to be handled with tact, as Quirós was to learn.

For his campmaster Mendaña made an extraordinary choice. Pedro Merino Manrique was a soldier of about 60, white haired, vigorous and belligerent, by reputation quick to draw his sword.

On coming on board at Callao he precipitated quarrels with the boatswain and with Quirós, which Mendaña had to settle. At Paita he had words with one of the priests and bickerings with Lorenzo Barreto, an altercation that promptly involved two soldiers, with Manrique drawing his sword. He then stormed ashore in a fury, sent for his belongings and had to be persuaded to re-embark. At this point Quirós, apparently wishing no part in a voyage riddled with conflict, also resigned, and noisy arguments involving several people continued on the wharf and on deck. Mendaña with 'honeyed words' pleaded with Quirós to return. Manrique, walking by, added his comments in a loud voice: 'Well, Sir! the Devil walks loose among us.' 'The two embarked', says the record, 'not very friendly'.[10]

SPE ET
METV.

Chapter Eight
'. . . WE OPENED FIRE
ON THEM'

15 89

Despite these quarrels, the voyage began well. The weather was fine and the south and south-southeast winds favourable as the four vessels carved their paths across a quiet sea. Hopes were high for new discoveries, riches, religious conversions and a new life in the colony. Fifteen marriages were solemnised.

Before sailing Mendaña had instructed Quirós to produce a map showing only the coast of Peru and two points in 7° and 12° South latitude at a distance of 1500 leagues west of Lima. Here the Solomon Islands would be found. By Mendaña's order, nothing else appeared on the chart lest one of his captains be tempted to veer off to seek some discovery of his own. Copies were distributed to the pilots of the other vessels. The maps later caused sharp criticism of Mendaña.

Thirty-five days after leaving Peru land was sighted. Although the land was just 1000 leagues from Peru, Mendaña joyfully assumed he had reached the Solomon Islands, and ordered the singing of the 'Te Deum Laudamus' as everyone knelt in thanksgiving. He named the island La Magdalena. Today it appears on the charts as Fatu Hiva, the southernmost of the Marquesas Islands. The following day the ships approached the island, where steep, rugged sides dropped from considerable heights to the sea. Seventy small dugout canoes with outriggers swarmed out to meet them, with some 400 men paddling or swimming alongside. They were naked, 'fair' or 'tawny', with graceful, 'well-formed' bodies and faces covered with patterns in blue, 'and so well built that they certainly had the advantage over us'.[1] Among them there was a child of about ten years of age. Later Quirós wrote, 'His eyes were fixed on the ship, and his countenance was like that of an angel . . . I never in my life felt such pain as when I thought that so fair a creature should be left to go to perdition.'[2] The memory of the beautiful child he saw as eternally damned never left him.

One native was coaxed to climb on board and Mendaña dressed him in a shirt and hat, which caused much laughter among his fellows. With that some 40 others clambered aboard, taking hold of things, touching the soldiers and handling their weapons. The varied colours of the Spaniards' clothing confused them, until some showed them bare chests, arms and legs. Given shirts, hats and other items, they danced and sang. Finally Mendaña signed that they should leave. When they refused and became aggressive in seizing various objects, Mendaña ordered a gun fired. Terrified, they leaped overboard and swam to their canoes. One man remained, clinging stubbornly to the main channels until someone wounded him in the hand with a sword and he scrambled down into a canoe. With this all friendship evaporated. The natives fastened a rope to the ship's bowsprit and by rowing tried to tow the vessel ashore. This failing, they rapped on seashells, beat with their paddles, hurled stones and brandished spears. 'It was a sight to behold,' said Quirós. The soldiers aimed their arquebuses but, damp from a recent rain, the powder failed at first to ignite. When it did, several natives were killed and others wounded. It was a presageful beginning.

In the following days the Spaniards came upon three more islands. Mendaña, realising that he was not among the Solomons but in an entirely different archipelago, named the group Las Marquesas de Mendoza, honouring the viceroy and marquis, García Hurtado de Mendoza.

Three days later Mendaña and most of the expedition's company, including Doña Isabel, came ashore to hear mass celebrated by the vicar. The natives present knelt as well, quietly following the motions made by the Christians. Sitting next to each other were Doña Isabel and a handsome native woman. In a story that exemplifies the discrepancies that can occur in different records of the same event, one version has Isabel so intrigued by the other woman's red hair that she wants a few locks cut off; in the other, the native woman wants some of Isabel's fair hair. In both versions, however, the request is declined.[3] Mendaña, meantime, took official possession of the four islands, visited the village, sowed some maize before the islanders and returned to the ship.

Subsequent encounters with the island people were intermittently friendly. Mendaña's chaplain, Antonio de Serpa, made a particular friend: 'they called one another comrade; he taught him to bless himself, and say *Jesus, Mary*'. The islander cheerfully

allowed himself to be taken aboard the *capitana*, which he explored
with great curiosity. There were other overtures. 'They asked of
one another by signs how they call'd the heaven, earth, sea, sun,
moon, and stars, and other things they saw'.[4] But with arquebuses
in the hands of some of Lima's rowdies, shooting erupted quickly
over trivialities. Told to fetch water in the ships' jars, one group
of natives ran off with four jars, 'for which reason we opened fire
on them'.[5]

In a boat with his soldiers, Manrique found himself
surrounded by men in canoes. To ensure their own safety, the
Spaniards 'killed some of them'. One man jumped into the sea
holding a child and was shot by a soldier, who afterwards said
regretfully that the Devil had to take those who were meant
to be taken. Quirós asked him why he had not fired into the
air. He replied that he had to maintain his reputation as a good
marksman. What would it 'serve him to enter into hell with the
fame of being a good shot?', Quirós asked.[6] Evidently there was
no reply. Manrique had free reign to exert his brutality, hanging
up the bodies of three slain warriors that their wounds might
be fully appreciated by their companions. Quirós recorded that
a 'certain person' came to see the bodies, giving one a lance
thrust and praising the deed. No word tells us who this was. 'At
night the natives took the bodies away.'[7]

Quirós believed that about 200 natives were killed in these
islands during the Spaniards' stay of under three weeks, possibly
but not necessarily an exaggeration. Perhaps trying to find some
rationality in the events that took place, he wrote:

> Our men were very well received by the natives, but it was not
> understood why they gave us a welcome, or what was their

intention. For we did not understand them; and to this may
be attributed the evil things that happened, which might have
been avoided if there had been some one to make us understand each other.[8]

The Spanish narrators of the expeditions recorded much
about the daily lives of the people they met. They described their
agricultural methods, house construction and village layout,
their dress or lack of it, boats, weapons and ornaments. They
also noted their appearance, especially the variations of skin
colouring that, although sometimes exaggerated, testified to the
diversity of racial strains to be found in the South Pacific. But of
the cultural or spiritual life of the people the Spaniards understood—and attempted to understand—very little. Their own
strong religious convictions and the evangelising mission on
which they saw themselves precluded any need to understand
mistaken beliefs which, in any case, they would soon supplant.

By the beginning of August the ships were wooded and
watered and ready for departure. The islands had been claimed
for King Philip and Mendaña considered leaving a group of
about 30 mostly married people as a colony. But angry opposition made him drop the idea. On 5 August he had three crosses
raised on the shore and a fourth cut on a tree, with the date.

The vessels weighed and in an east wind steered to the
west-southwest. Three or four days and some 400 leagues later
Mendaña announced that they would see land on that very day.
The result was an outburst of joy and relief, food and water
happily and liberally consumed, and an eager lookout kept by
almost everyone. No land appeared. On the 20th four low-lying,
wooded but seemingly uninhabited islets were sighted in, by

thcir calculations, 10° 20' South latitude. Mendaña named them San Bernardo but, seemingly after some hesitation, made no attempt to land. Apparently Quirós favoured a landing, for he later commented in an account written for Antonio de Morga, Lieutenant-Governor of the Philippines, that 'indeed that year seemed to be the year of the pusillanimous, though it makes me angry to have to say so'.[9]

Nine days later there appeared a low, round little island, thickly grown with coconut palms and bushes, but wreathed with reefs, which Mendaña named La Solitaria, now known as Niulakita in Tuvalu. With water and firewood sorely needed, particularly by the *almiranta*, Mendaña sent sailors in two boats to search for a landing site. They found an uneven sea bottom and many 'great rocks' and with much shouting warned off the ships. The fleet moved on. Quirós placed the island at 10° 40' S and 1535 leagues from Lima. No substantial improvements in navigational instruments or techniques had been made since the voyage piloted by Hernando Gallego 25 years before, and Quirós' results in this respect were much the same. Like Gallego, he largely underestimated the distances sailed, although the error was usually somewhat less, which could be accounted for by his more southerly route which reduced the effect of the westward-flowing South Equatorial Current.

With the feasting that had taken place when Mendaña announced that land was shortly expected, water and provisions were running low. Cynical but alarming rumours circulated. The Isles of Solomon had been passed unseen or had been swamped by a rising sea. Mendaña had forgotten their location. Or they would be sailing on forever or at least to Great Tartary. Mendaña countered by ordering prayers to be said before the image of the Virgin and sacred banners displayed. He himself had his rosary in his hand at all times.

Thirty days after Mendaña had promised the immediate sight of land, a dark cloud was seen on the horizon. As they approached it grew into a great, turbid, rising mass of black smoke. Quirós ordered the frigate and the galeot to go ahead, keeping sight of each other and signalling by torchlight as night came on. At nine o'clock the *capitana* and the *almiranta* were sailing in company. At eleven, heavy rain clouds obscured the horizon, but when they lifted the dark outline of land could be seen less than a league away. The galeot continued to signal, points of light in the darkness, and two vessels responded. There was no response from the third.

As the grey light of dawn crept into the sky, they saw 'a single pointed mountain rising out of the sea', bare slopes deeply scored with crevices and a ragged shoreline dropping steeply into the water. They had come upon the volcano-island of Tinakula, an almost perfect 800-metre cone.[10] The *almiranta*, the *Santa Isabel*, was nowhere to be seen. Deeply alarmed, Mendaña sent the frigate to sail around the volcano in search of it and, backed by the vicar, ordered the people on board to be confessed, confessing in public himself.

Ahead of them was the land they had seen the night before, large, mountainous, densely wooded right down to the sea. A fleet of small canoes emerged, the men in them black-skinned, some 'tawny', with 'frizzled' hair, waving and calling to the ships. Mendaña immediately took them for Solomon Islanders and spoke to them in what he knew of the languages of San Cristóbal and Guadalcanal. They did not understand him. The canoes milled indecisively about the ships, the men chattering, until suddenly, with a great shout, they shot their arrows into the sails and onto the decks. Their arquebuses prepared, soldiers opened fire. Quirós recorded, 'Some were killed, many others wounded, and they all fled in great terror'.[11]

The *capitana* now stood off and on, while the galeot searched the shore for a suitable harbour. The frigate returned. There had been no sighting of the *Santa Isabel*. Mariana de Castro, newly wedded wife of the *almiranta*'s captain, Lope de Vega, collapsed in tears. She had remained on the *capitana* with her sister Isabel. Others too were stricken. There had been some 130 people, many of them friends and relatives, on the *Santa Isabel*. Mendaña's thoughts were reportedly bitter. Apparently he believed that the *almiranta* had purposefully deserted the squadron, and her contingent of ten soldiers and their armament was a serious loss. Years later Juan de Iturbe, an accountant on Quirós' 1605–06 voyage, claimed that people on Taumako (Taumaco) in the Duff Islands gave the Spaniards to understand 'by signs' that the men of the *Santa Isabel* had been slain, but the women and children had survived at a location only 60 leagues away. No other report supports this claim.

The disappearance of the *Santa Isabel* remained a mystery for 375 years. In 1970–71, archaeological excavations at Pamua on the coast of the Solomon Islands' San Cristóbal uncovered what was evidently Spanish or Spanish colonial pottery in quantities that exceeded anything that could have been left in 1568 by Hernando Gallego's brigantine crew, who had explored the site but never camped there. Specimens had been found in the area since the 1920s, but the more recent investigation firmed their identification. On the only hill in the vicinity, with a view of the sea and a beach and with fresh water and flat, fertile land below, two fireplaces were carbon-dated to the late 16th century, and hundreds of sherds and a scattering of metal artefacts uncovered. That a group of Spaniards had lived here was evident. It seems reasonable to conjecture that the *Santa Isabel* found the Solomons, the goal of the expedition, and waited for the others at San Cristóbal. But why or how the *almiranta* had separated from the fleet remains an unanswered question, and what eventually became of the ship and its company of some 130 men, women and boys is a mystery with most likely a tragic ending.[12]

Four days later Mendaña's fleet found a better port. The ships anchored in 15 fathoms within a deep bay, sheltered from the wind and near a native village and a river. Quirós wrote, 'All night the noise of music and dancing was heard, striking drums and tambourines of hollow wood'.[13] It would have been an eerie night, strange sounds throbbing across the water in the warm, humid darkness. Where were they? Mendaña soon realised that they were not among the islands he had sought, and he named his new discovery Santa Cruz. It was, in fact, the island of Ndeni (Ndendo or Nendo) in the group still called Santa Cruz, today part of the Solomons and roughly about 400 kilometres from San Cristóbal. The inexact coordinates given for the Solomons during the first voyage, together with an underestimate of the distance they had just sailed, had placed the expedition well short of their destination.

To Mendaña the location evidently seemed as suitable a jumping-off point for the continued search for Terra Australis as any of the Solomons.

Now the volcano of Tinakula erupted. Sparks and fire poured from the summit and from the crevices along its sides, with tremendous roars, and tremors that rocked the ships anchored 10 leagues away. Smoke seemed to cover the 'whole concave' of the sky. An awakened Tinakula was an unsettling neighbour.

Initially the meeting with the local people was congenial, even sociable. Islanders flocked to see the ships, many with 'red flowers in their hair and in their nostrils'.[14] Leaving their weapons in the canoes, they boarded the *San Gerónimo* where their chief, Malope, an impressive-looking man, received gifts and in the accepted gesture of friendship exchanged names with Mendaña, while soldiers and crew handed out little bells, feathers, glass beads, pieces of cloth, even playing cards. They showed their visitors mirrors, shaved their heads and chins with razors and cut their finger and toe nails with scissors, and let them know 'what was under the men's clothes', all with great hilarity. Lorenzo Barreto, meanwhile, left in the frigate with twenty soldiers and seamen to search again for the *Santa Isabel*. They found no sign.

Peaceful relations did not last. What followed was a confused welter of warfare among native groups, attacked and injured Spaniards, ferocious revenge by Manrique, restored friendships and new misunderstandings. The Spaniards shifted their vessels to yet another location, where there were further battles and a quarrel between Manrique and Lorenzo Barreto. Here, for the first time, Isabel Barreto's interference is recorded. Her brother, she said, had unlimited authority in military affairs. Apparently she spoke insultingly of Manrique, who angrily spent the night ashore in a native village.

Establishing the colony as a base for further exploration was now essential. The harbour finally chosen by the Spaniards was a bay on the island's north side, about 5.6 kilometres in length.

Mendaña called it Bahía Graciosa, Graciosa Bay, a name that still appears on the charts. Ashore there was a stream of clear water and a river beyond. A quarrel promptly broke out between Manrique and Mendaña on the location of the settlement, the unmarried men backing the campmaster's choice, the married men maintaining that the site was unhealthy and proposing to occupy a native village where the houses were ready-built and the ground beaten smooth. A compromise was reached, and the soldiers set to work felling trees and erecting huts with palm leaf roofs. In Quirós' words, they worked with 'good will' for God and their king. It would have been hard, hot, sweaty work, at times in the drenching rain of thunderstorms. As the huts were constructed, soldiers moved from the ship to the encampment.

Despite the willing start, however, Quirós soon noted that there were those with whom, as he wrote, the Devil was at work. Many resented Mendaña's orders to respect the natives and their property, believing that with all they had left behind and suffered to come to this coast, everything here should belong to them. For a time, however, the work proceeded without incident. Archaeological evidence of the settlement was uncovered in 1970, when excavations revealed a house floor, a stone alignment, two burials, and 120 pottery sherds identifiable as of 16th century Spanish or Spanish colonial manufacture.[15]

Food, fortunately, was not yet an issue. The local people brought provisions—nuts, fruit, including the first recorded breadfruit, and various roots. Soldiers returned from evidently peaceful bartering expeditions with coconuts, plantains and pigs. Good relations seemed sufficiently established for Lorenzo Barreto to reach an agreement with the villagers: they would assist in building the settlement, and no house of theirs would be damaged. The time for evangelisation seemed to have come. The vicar made a cross of two poles and in a procession carried it through the native village. A church was built and a priest said mass daily.

Nevertheless an undercurrent of rebellion was running through the Spanish camp. Dissatisfied with almost everything, and with sickness, probably malaria, spreading among them, an increasing number of men wanted to find a better place. Someone drew up a petition with several signatures expressing this sentiment, and it was taken to Mendaña. The *adelantado* had continued to live on the ship, his house not being ready. He was unwell, but now went ashore to meet some of the ringleaders. Nothing came of the meeting. Mendaña returned to the ship, sending Jordán Estomate (Estomase), pilot of the galeot, to deal with the protest. Apparently ignoring any instructions from Mendaña, Estomate encouraged the men to leave, promising to take them to a better island or to Peru. Now Lima's undesirables came up with a plan to kill enough natives to arouse them to war. With that would come a scarcity of food and the necessity of leaving the island. The men vacillated, quarrelled and even terrorised each other.

Mendaña made additional visits ashore. Quirós commented that he showed 'great patience, and . . . much sorrow' as he listened to the men's grievances, but did nothing. Marking out the site for a stockade, allotting land for cultivation and discussing the future administration of the settlement brought on a storm of demands and arguments on ownership, land titles and entailments. Someone fired shots at the *capitana*. The Barreto brothers believed their lives were in danger. The vicar, going ashore to say mass, returned to the ship convinced that the men would compel a departure from the island.

Little could be expected from Mendaña, ill and now deeply despondent, but Quirós received permission to speak to the malcontents, among whom he evidently had some personal friends. In the camp he was crowded about by the men, pelted with questions, threats and demands. Later he wrote:

> Some were saying:'Where have you brought us to? What place is this where no man goes . . . people would only come to take gold, silver, pearls, or other things of value, and these are not here.'
>
> Others said: 'We did not come to sow: for that purpose there is plenty of land in Peru . . . Embark us and take us to seek those other islands, or take us to Peru or some part where there are Christians.'[16]

Quirós reminded them that the cities of the world—Seville, Rome, Venice—were once 'forests or bare plains'. Abandoning their enterprise made them 'enemies to God . . . the King . . . the honour of our General . . . our own honour'. 'I was without a sword', Quirós wrote, 'and he [one of the soldiers] with seven or eight others, went for theirs . . . God knows what they intended'. Somehow tempers cooled. Quirós returned to the ship, but nothing had changed. As he would show again, Quirós would not step beyond the limits of his position as chief pilot. In the midst of seething hostility, he remained cautious and restrained, almost aloof.

The campmaster, however, was treading a narrow line between the opposing groups, which was obvious to Mendaña and his wife. 'Kill him or have him killed,' Isabel urged her husband when Manrique came on board to pledge his loyalty, 'if not I will kill him with this knife.'[17] Mendaña, despite his own anger, restrained her, and Manrique returned to the camp, assuring the soldiers that although he served his general and his king, he favoured their side. Mendaña, meantime, quietly arranged with Lorenzo Barreto to have the campmaster put to death. Quirós later commented, 'This was very different from what I had understood that he intended to do, but . . . let those of better understanding judge.'[18] His loyalty to his superiors remained intact.

Quirós then received Mendaña's permission to look for food. He was met by Malope, the chief who had initially befriended the Spaniards and who now demonstrated his friendship further, going from village to village to collect edibles. Challenged by some of his own people, Malope hesitated but, threatened by Quirós, continued to seek supplies. At his order more than a hundred men shouldered foodstuffs and carried them to the ship's boat.

But no peace resulted. The next day Mendaña landed with his group of armed men, among them Felipe Corzo, the galeot captain, carrying a large 'wood knife'. Joined by Lorenzo Barreto, his brothers and a few sailors, they went to the fort being constructed by the campmaster. Manrique, half-dressed, was summoned from his breakfast. According to Quirós, Mendaña 'raised his eyes to heaven, and, giving a sigh, put his hand to his sword, saying, "Long live the King! Death to traitors!"'. With that the slaughter began. Manrique was cut down, crying 'Jesus Maria!', while his friends and servants were pursued and slain. Diego Barreto hoisted the royal standard and cried out for the king, those around him responding hysterically, 'Death to traitors!' Quirós shielded those he could from the swords of Luis and Lorenzo Barreto. 'The men were like lunatics', he wrote, 'going about with their eyes seeking those they would kill, shouting with drawn swords . . . this was a day for avenging injuries'.[19]

Amid the clashing of weapons and the screams and shouting, the women cowered in terror, weeping, praying for their men, while the drummer hurriedly stripped Manrique's body of clothes he wanted for himself. A boat 'in a great hurry' arrived from the ship, with the vicar holding a lance and armed sailors shouting 'Long live the king! Death to traitors!' as they sprang onto the beach and rushed to support Mendaña. The fighting, however, was finished.

Two severed heads were put into nets and Corzo set each on a pole near the settlement's guard post. On being told of the victory, Doña Isabel and her sister arrived from the ship and Isabel went promptly to view the heads. Mendaña ordered everyone to go to church, where the vicar said mass before the exhausted, bloodstained crowd. The deaths, he told them, had been ordained, and safety lay with being obedient to the *adelantado*. The belongings of the dead were shared among their killers, and the bodies buried.

Almost forgotten in the terror of the day was a squad of 30 soldiers who had left early in the morning to get more food from the generous Malope, whom they were warned not to harm. Laughing, they went off. Now in the afternoon one of the group arrived to report to Mendaña. Malope had invited them to a feast and handed over what food he had left. Then one of the arquebusiers raised his gun and shot him. Another soldier completed the murder by cleaving Malope's head with a hatchet.

As the soldiers came marching back in the afternoon, unaware of events in the camp, Mendaña had them seized and put in the stocks. The ensign in charge was beheaded and his body thrown into the sea. His wife wept bitterly. The arquebusier who had shot Malope was to be executed next, but Quirós interceded, pointing out that in present circumstances every man was needed. The soldier was sent on board under guard, but died suddenly a few days later. Others took an oath of loyalty to Mendaña and lived.

The next morning Mendaña ordered a party to take the head of the ensign to Malope's village, hoping that the natives would believe this to be the head of the murderer. The people fled and even with the head held up to be seen, could not be called back. It was then placed before the doorway of Malope's house. The two heads set on poles at the guard post were taken down, but left on the beach, and were mostly eaten overnight by the dogs. Mendaña, now seriously ill, ordered his thatch house to be completed quickly and moved there with his family. Bedridden, he was vomiting blood. Lorenzo Barreto was put in charge of the colony.

Malope's infuriated people attacked, assailing the palisade around the camp with arrows, stones and great shouting. Lorenzo, carrying the banner of Spain, led a charge in defence of the gate. The native warriors fell back but fired a volley of arrows that wounded six Spaniards, including Lorenzo. The soldiers then rushed the native village, setting alight houses and canoes, but suffering another eight injured. The islanders had learned to fire their arrows at the soldiers' legs and eyes, unprotected by shields, helmets or other armour. Emboldened, they began attacking at night. One Spaniard was injured, another killed. Casualties from illness also mounted. Mendaña's chaplain, Antonio de Serpa, died.

On the night of 17 October there occurred the alarming event of a total eclipse of the moon. In his house by the light of an oil lamp Mendaña dictated his will, which he was almost too weak to sign. He named his wife, Isabel Barreto, as his heir and governor of the colony and bequeathed to her 'all his property, rights and titles and the same to whomsoever she should marry'.[20] He nominated his brother-in-law Lorenzo Barreto as captain-general. The vicar administered the final rites. At one o'clock in the afternoon of 18 October 1595, Alvaro de Mendaña de Neira died. He was carried to the church with all the pomp the little community could muster, in a black-draped coffin borne by the expedition's highest ranking men, soldiers lining the path with their arquebuses reversed, banners displayed, a slow and muffled drumbeat, the mournful piping of a fife. Mendaña's will was executed the next day and Doña Isabel became the colony's governor.

Now two or three people were dying each day. The vicar walked through the camp, calling out loudly, 'It there one who wants to confess?' God's 'terrible chastisement' was upon them for their many sins if they did not do so.[21] But soon after this the vicar, too, fell ill, and returned to the ship. Quirós debated the cause of the sickness, observing that those who remained 'on the sea' did not become ill, but no one made the connection between the fever and the sting of a mosquito. Probably, too, even minor wounds developed into fatal infections. There were neither remedies nor medical men with the expedition, not that either, with the limited knowledge of the time, would have saved very many.

Despite his own failing health, Lorenzo tried to exert control. Together with Quirós he made a bid for peace with Malope's people, assuring them that those who had slain their chief were dead. For the Spaniards peace was essential. Unable to leave the encampment in search of food, they had been surviving on flour brought from Peru. The islanders responded with remarkable goodwill, coming with provisions the following day and again later. Lorenzo sent the frigate on one more attempt to find the *Santa Isabel*, but again no trace of the ship was found. The purpose of the settlement seemed to be dying with its people. On 30 October the vicar, officially representing the soldiers, drew up a petition listing reasons for abandoning the moribund colony. The document was carefully written, signed or witnessed by virtually everyone. The next day Lorenzo Barreto made the final decision for departure, which Isabel Barreto authorised.

Maris Pacifici, map of the Pacific Ocean (1601) by Abraham Ortelius. New Guinea is shown as an island, but this was purely speculation. Terra Australis occupies much of the southern hemisphere. (National Library of Australia)

Pope Clement VII (1536–1605), engraving by Holman. Clement supported the plans of Quirós to create a Spanish colony of Terra Australis and to introduce Catholicism to its imagined population. (Mary Evans Picture Library)

Philip III (1578–1621), King of Spain and Portugal, engraving by Goya after a painting by Velasquez. In 1603, Philip provided Quirós with orders to the viceroy of Peru to organise a well-equipped expedition for the discovery of Terra Australis. (Mary Evans Picture Library)

145. Philipp III., König von Spanie... Nach einem Gemälde von Diego Velasquez gestochen von Fr. Goya.

Map of the Bay of San Felipe y Santiago on the island of Espíritu Santo (1606), now part of Vanuatu, by Diego de Prado y Tovar. Small anchors show the ships' anchorage site. Geographical features were named for members of the ships' companies. (España. Ministerio de Cultura. Archivo General de Simancas. MPD, 8, 82)

Nagheer Island, formerly Mount Ernest. The men of Torres' expedition named it Isla de Nuestra Señora de Montserrat for its rugged appearance reminiscent of the massive mountain of Montserrat near Barcelona. (Photo by A.K. Estensen)

Torres' ship the *San Pedro* and the launch *Los Tres Reyes* sailing within sight of Cape York, Australia, an imagined depiction by J.R. Ashton. (From a copy of the *Picturesque Atlas of Australasia*, Andrew Garran ed., held by the Mitchell Library of New South Wales, Sydney)

An 18th century engraving of the Spice Island of Ternate shows the settlement under Dutch control. Busy trade in spices, especially cloves, is in progress. The island was in Spanish hands in 1606 when visited by Torres sailing from New Guinea to the Philippines. (Courtesy of Dr Nanne Sjerp)

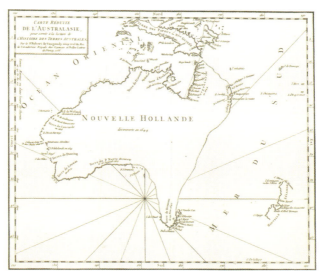

Carte reduite de l'Australasie (1756), here named Nouvelle Hollande, by Gilles Robert de Vaugondy. Quirós' 1606 landfall of Espíritu Santo (Vanuatu) was believed by some later geographers to be on the unknown east coast of Australia. Quirós' New Jerusalem and Jordan River are shown. Parts of Tasmania and New Zealand were known through the explorations of Abel Tasman. (National Library of Australia)

Chapter Nine
JOURNEY TO MANILA

Lorenzo was dying. Apparently an infection in the arrow wound he had received in the leg was taking his life. He lay with his eyes fixed on the crucifix, resigned to dying without being confessed. 'Ah, death! in what a condition you take me . . . I am a sinner. O, Lord! pardon me.' Determined to enable Lorenzo to die confessed, Quirós went aboard the *capitana* and pleaded with the equally infirm vicar to come to him. The vicar

> was put in the boat, trembling and wrapped in a blanket. He was carried to the side of Don Lorenzo in his bed, whom he confessed, as well as all others who wished to confess. A soldier, seeing how ill the Vicar was, said very sorrowfully: 'Ah, sir! what is this I see? What have we come to?' They returned to the ship.[1]

At daybreak on 2 November Lorenzo died. Quirós wrote: 'Our condition . . . had reached such a point that, if only ten determined natives had come, they could have killed us all, and destroyed the settlement'.[2]

Now the sick were taken on board and Doña Isabel also left the encampment. On Monday 7 November, the flag and the last of the Spaniards were embarked. The dogs, left behind, ran barking along the beach. Then the smallest one rushed into the water and swam out to the galleon. Such fidelity could not be ignored, and he was scooped up and taken aboard. The next day the vicar, Juan Rodríguez de Espinosa, died and was buried at sea. 'This loss was what we deserved for our sins', wrote Quirós.[3]

Two final forays were made in the boat to seize provisions for the long voyage ahead; despite contrary winds and brief skirmishes with angry natives, the boat returned well laden each time.

Doña Isabel, as governor, directed that the fleet should sail first in search of San Cristóbal and the *almiranta*. If neither were found, the expedition would head for Manila. Quirós pointed out that the three vessels were dangerously worn and seriously undermanned, and suggested that the galeot and the frigate be abandoned and their rigging, sails and people put to use aboard the *capitana*. Argument ensued and nothing was done. Someone proposed that the sick and wounded be transferred to the frigate, where only a tent would protect them from the weather, but here Quirós prevailed upon Isabel to order that they remain on board the *capitana*. That night Diego de la Vera, the captain of the frigate, and some of his men slipped quietly ashore, disinterred Mendaña's body and placed it in their craft to be taken to Manila.

On 18 November 1595 the three vessels sailed from Bahía Graciosa. Forty-seven people had died there in the space of a month. Quirós noted that the survivors 'turned their eyes to the huts of the settlement, saying, "Ah! there you remain, thou corner of Hell, that cost us so much!".'[4]

The vessels steered west-southwest. At 11° South they would have been tantalisingly close to San Cristóbal, where very likely the people of the *Santa Isabel* were encamped and waiting. On the *capitana*, however, the boatswain and four more seamen had fallen ill, and the soldiers and crew protested against spending more time in the search. Doña Isabel agreed to change course for Manila, which Quirós calculated to be about 900 leagues from Bahía Graciosa. Accordingly, he shifted his heading to the northwest.

Squalls and calms came and went, bludgeoning the fragile vessels to the terror of the people, or baking them in merciless sunlight as they lay almost motionless on a flat, dazzling sea. The daily ration was half a pound of flour per person, mixed with water into a paste that was baked in hot ashes, and a pint of water stinking with disintegrating cockroaches. The strongest were given double rations to enable them to work at the pumps, which they had to be driven to do.

People continued to die, the living so weak that they could scarcely bring the dead up from the lower decks. Bodies went overboard almost daily, sometimes several times a day. Sewing and splicing were continuous on the rotting sails and rigging. The mainmast of the *capitana* was sprung from its step. The spritsail fell into the sea and could not be retrieved. Topsails and the mizzen sail were taken down to mend the courses, which became the only sails left, while a halyard was spliced 33 times and broke again. Throughout, Doña Isabel remained ensconced in her cabin, consuming her private store of provisions—oil, wine, biscuits, fresh meat, water that she used for washing her clothes. Quirós pleaded for a share for the men. Her reply was that if two of them were hanged, the rest would cease their importuning. Grudgingly, she handed over two jars of oil, which were soon consumed.

Early in December the galeot disappeared. Although its captain, Felipe Corzo, had been ordered to maintain a position not more than half a league from the *capitana*, should the larger vessel sink, the galeot stood on another tack one night and vanished into the darkness. The frigate, meantime, was leaking badly. Doña Isabel refused to order it abandoned and its captain would not leave his craft, which carried Mendaña's body. Yet a morning came when the frigate was not to be seen. Quirós had the *capitana* hove-to until the following afternoon, when the outcry from the soldiers against precious time being wasted forced him to get under sail again.

On 3 January 1596 they sighted Guam, the southernmost of the Mariana Islands, and Saipan, in the group Quirós knew as the Ladrones, named by Magellan in 1521 and formally claimed by Spain in 1565. A seaman handing the foresail fell overboard: 'in the whole ship there was only one line', which was thrown to him and which he caught.[5] 'Thanks be to God', wrote Quirós.

Canoes under sail converged on the ship, the people shouting for pieces of iron. Plantains, coconuts, rice, some large fish and water were handed up in exchange. To go ashore for more provisions was not possible, 'for we had no cable for lowering and hauling the boat'.[6] The trading, however, went well except for the killing of two natives 'by an arquebus, owing to a matter of a piece of cask hoop'.[7]

A brief comment on this final meeting with an island people exists in a piece of writing in which the unnamed author looks back upon a primitive man's existence, where with his small, light canoe and habitation in a tree, he lives 'barbarously, yet happy in being a stranger to the fall of court favourites, and to the favours of the world, such as wealth, honour, and preferments, imaginary blessings and chimerical delights'.[8] The disillusioned and world-weary voice was probably that of Quirós.

Without a chart Quirós could only steer west in search of the Philippines, and at daybreak on 14 January a mountain peak was sighted. Along the island of Samar they coasted, seeking a safe anchorage for the ruined ship. In the Bay of Cobos they were met by friendly natives who exchanged quantities of food for coins, knives and glass beads. 'During three days and nights the galley fire was never put out, nor was there any cessation of kneading and cooking or of eating the boiled of one and the roast of another, so they were eating day and night.'[9] Of the sick, however, two or three died. Quirós, as chief pilot the object of so much abuse in times of suffering, was now the recipient of overwhelming gratitude. Their thanks belonged to God, he told everyone.

The *San Gerónimo* sailed again on 29 January,[10] the crew committing two more bodies to the sea as they left. Two days later the boat was somehow launched and the governor sent her brothers Diego and Luis ashore to look for food. They did not return, having decided to go on to Manila by a shorter, partially overland route. In the following week, moving slowly among the islands in a ship with only two working sails, food and water again ran out. Once more Quirós found himself caught between the ruthless parsimony of Doña Isabel, who refused to surrender any of her private store of provisions, and the desperation of hungry, menacing soldiers and seamen. Reluctantly Isabel gave up a calf. In Quirós' words, 'there were many painful scenes'.

At the entrance to Manila Bay, Spanish officers and soldiers boarded, received with frantic joy by the men on deck, 'sick, covered with boils, poverty-stricken, with tattered clothes . . . and between decks . . . the sick women', crying out for food and water. A horrified official saw Doña Isabel's two pigs and demanded to know why they had not been slaughtered. 'What the Devil!' he said to her. 'Is this a time for courtesy with pigs?' Isabel ordered the pigs to be butchered.[11]

A second boat brought Manila's chief magistrate and Diego and Luis Barreto, and the next day a large barge arrived with food and wine, while sailors from a royal vessel came on board to assist the *San Gerónimo*'s debilitated crewmen. On 11 February 1596 the ship let go her anchor in Manila's inner port of Cavite, as on the shore the banner of Spain flew in the wind and cannon and drawn-up arquebusiers fired a salute. 'The ship replied as well as she could.'[12]

Fifty people had died on the twelve-week voyage from Santa Cruz to Manila. Of the approximately 378 who had sailed from Peru, about a hundred remained. Ten more died after arrival in Manila.

The galeot reached the southern island of Mindanao after the starving crew had landed on an islet to kill and eat a dog they saw on the beach. Guided by natives to some Jesuits and subsequently to the local Spanish governor, the survivors, including their captain, Felipe Corzo, were sent on to Manila, perhaps by outrigger craft, boats commonly used for inter island transport. No explanation of Corzo's separating from the fleet seems to exist, other than his known hostility towards Quirós. Of the frigate carrying Mendaña's body it was rumoured that it had been seen aground on some unknown beach, sails set and all on board dead.

Antonio de Morga, Lieutenant-General of the Philippines, ordered an investigation of the voyage, with copies of all relevant documents to be sent to the king. Quirós' official account was relatively brief on events on Santa Cruz, but Isabel Barreto acquitted him of any questionable act, commending him for his service and unwavering loyalty to the king and to her late husband. Other witnesses testified similarly.

Doña Isabel was honoured and feted in Manila, which was anticipating with festivals the arrival of a new governor, Don Francisco Tello de Guzman. The entertainments included performances by three elephants, gifts from Southeast Asian princes to previous governors, which absolutely fascinated Quirós. In May 1596, only three months after reaching Manila, Isabel married Fernando de Castro, a nephew or by some accounts a cousin of a previous governor and a knight of the Order of Santiago. 'As was just', in Quirós' words, Castro took possession of her inheritance, the property, rights, titles and responsibilities originally granted by the king to Mendaña for the settlement of the Solomon Islands.

Fernando de Castro refitted the *San Gerónimo*, took on a cargo of valuable Chinese merchandise, and with his wife and Quirós sailed for Mexico. It was a long, stormy and terrify-ing journey, but on 11 December 1596 the weather-beaten ship struggled into Acapulco. In an investigation before local authorities of the outcome of Mendaña's expedition, Quirós was absolved of any responsibility for the killings on Santa Cruz. 'There I, Captain Pedro Fernández de Quirós, took leave of the Governoress and of my other companions, and embarked on board a passenger ship for Peru.'[13]

For many years Isabel and her new husband tried unsuccessfully to obtain permission to establish a settlement in the Solomon Islands, their inheritance from Alvaro de Mendaña and potentially the springboard for the discovery of Terra Australis. On 15 July 1612, Isabel Barreto, the woman who had largely financed the voyage to Santa Cruz and briefly governed a Spanish Pacific island colony, died at Castrovirreyna, Peru.

SPE ET METV.

Chapter Ten

QUIRÓS

On 3 May 1597, Pedro Fernández de Quirós came ashore at the Peruvian port of Paita, which he had left almost exactly two years before. He wrote:

> Thence I wrote a letter to the Viceroy, Don Luis de Velasco, and travelled by land to Lima, where I arrived on the 5th of June, and was well received by the Viceroy. He desired to be informed respecting all the particulars of our voyage and discoveries, and I gave him the best account in my power.

Then Quirós went on to what clearly had occupied his thinking for many months.

> I also offered that, if he would give me a vessel of 70 tons and

40 sailors, I would return to discover those lands and many others which I suspect to exist, and even felt certain that I should find in those seas.[1]

Quirós had taken on in its entirety the dream of Alvaro de Mendaña de Neira, a road he too would follow to the end of his life. Obdurate hope, almost endless disappointment, desperate poverty and moments of impassioned triumph would mark that road. Who then was this man Quirós, who now burdened himself with another man's insubstantial vision, with the pursuit of the mythical continent of Terra Australis Incognita?

Quirós was born in the ancient walled city of Evora on the arid plateau region of Alentejo in south-central Portugal. No baptismal record seems to exist, but a later reference to his age suggests that he was born in 1565. Evora had once been a Roman military base, was conquered about 712 by the Moors and retaken by Christians in 1166. The boy Pedro would have walked along its narrow streets, looked upon its Roman ruins, and perhaps peered into the dim, candle-lit interior of the cathedral rebuilt around 1400 in Gothic style. The great navigator Vasco da Gama had married and lived in Evora in the early 1500s, rich and honoured for the voyages that established Portuguese mercantile hegemony in India and today's Indonesia. Without doubt young Pedro heard many tales of his exploits.

In 1580 Spain annexed Portugal. Following the death of a young king and a succession crisis, the armies of King Philip of Spain overwhelmed the smaller nation and on 12 September the Spanish monarch was proclaimed King Philip I of Portugal. Quirós was about fifteen years old. Three years later King Philip paid a state visit to Evora, and quite possibly the youth witnessed the colour and glitter and heard the cadence of drums and trumpets as the procession wound through the narrow streets. A letter written many years later suggested that Quirós came from a disreputable district of Lisbon.[2] The author was, however, a bitter enemy of Quirós and the phrasing of the remark suggests that it was vituperative rather than factual. Quirós always referred to himself as being 'of Evora'.

Later Spanish documents generally referred to him as *de nación portuguesa*, of Portuguese nationality, but clearly he accepted a Spanish ruler without demur. He wrote of Philip as 'my natural King and Lord'[3] and adopted the Spanish spelling of his name.[4] Years later he justified some of his actions with a reference to the 4th century Roman emperor Theodosius I (c.346–395), the 15th century Albanian hero George Kastrioti, known as Skanderberg (1405–68), and the Holy Roman Emperor Charles V and King of Spain Charles I (1500–58), all men who had strongly espoused Christianity. These and many other 'valorous and prudent Captains', he wrote, were mirrors into which he looked day and night with the desire of imitating them.[5] The sense of an exalted religious mission entwines these thoughts.

When and where Quirós first went to sea is not known, but the letter referring to Lisbon states that he embarked as a clerk on a Portuguese merchant ship.[6] If so, he evidently went on to serve a pilot's apprenticeship, perhaps on coastal vessels visiting Iberian ports, and eventually underwent the examination for pilots at Seville's Casa de Contratación. At some point he crossed the Atlantic, perhaps as a pilot with one of the convoys plying between Spain and her American colonies. Probably in 1589 Quirós married Ana Chacón, daughter of a Madrid licentiate. A son, Francisco, and a daughter, Jerónima, were born, and later stated to be *naturales de la villa de Madrid*, that is, natives of the town of Madrid. Their dates of birth, however, are unclear. Curiously, Quirós never mentioned his wife and children in his voluminous writings, until in 1615, returning to Peru from Spain, he asked that his family be permitted to accompany him. Of Quirós' physical appearance the only surviving remark occurs in a document written when he was in his fifties: he had a bronzed complexion and a mole on his right nostril.

Quirós first comes into the light of history in 1594, when he is recorded as having interrogated at Lima an English geographer captured after the sea battle off Callao in which the wounded English captain Richard Hawkins was forced to surrender. The geographer told Quirós that the land called Tierra del Fuego, at the foot of South America, was an island with open sea beyond, a claim Quirós dismissed as intended misinformation.

In the following year, 1595, Quirós, obviously regarded as a highly qualified seaman and pilot, was asked to sail with Mendaña as a ship's captain and the chief pilot of the expedition intended to create a Spanish colony in the Solomon Islands. He accepted, but repelled by the disagreements on board, soon resigned. Persuaded to retract his resignation, on 16 June 1595 he sailed with the expedition. Thus it was that Quirós witnessed the debacle of the Spanish settlement at Santa Cruz, and with consummate seamanship and navigational skill brought the survivors in the foundering *San Gerónimo* across an unknown, uncharted sea to Manila.

The man returning to the Peruvian viceroyalty in 1597 was not, however, quite the same man who had left it two years earlier, for at some point during that catastrophic voyage the mind and heart of Pedro Fernández de Quirós had moved from what seems like

an almost casual interest in Terra Australis to a profound commitment to its discovery. An early inference of this can be found in a
personal letter he wrote in Manila in 1596 to Antonio de Morga,
Governor of the Philippines. Here Quirós asked that the discovery of the Marquesas Islands be kept secret until 'his Majesty be
informed and order what is most convenient for his service'. The
islands lying roughly midway between South America and the
Philippines, 'the English could do much harm in these waters if
they got to hear of them and settle there'.[7] He added, 'for man
does not know what time brings'.[8]

There was logic in this request, but Quirós' principal motive
was probably otherwise. Ostensibly a warning against English
aggression, the letter seems also to reflect the wish of a discoverer to keep his findings to himself until he could further
deal with them. A chart dated 1598, which is apparently the
only extant map drawn by Quirós, demonstrates this preoccu-
pation with secrecy, for it depicts none of the discoveries of
1595. Perhaps they were already in his mind the stepping stones
to the much greater discovery of Terra Australis.

Now, after landing at Paita, Quirós travelled to Lima to
be received by the viceroy, Don Luis de Velasco, Marquis of
Salinas. His first task would have been to report on the terrible
failure of the expedition. There were also navigational failures
to explain. Quirós as chief pilot had not found Mendaña's
Solomon Islands. This, Quirós explained, was because of errors
in observed latitude and because 'they had less longitude' than
they had thought, due to the mistakes of Hernando Gallego,
chief pilot on Mendaña's first expedition. Nevertheless, Quirós
believed that the Solomon Islands, Santa Cruz and New Guinea
were all in close proximity. His real interest, however, was in

the four islands which Mendaña had named Las Marquesas de Mendoza, the first South Pacific islands seen by Quirós. He had found their idyllic tropical beauty and the splendid physical appearance of their Polynesian people unforgettable. Here he had seen the child with a face 'like that of an angel' doomed, as he believed, to perdition. These were people that had to be guided to God.

The Spanish had no understanding of the sailing skills of the Polynesians and, seeing neither large ships nor navigational instruments, assumed that the people could only have come from a nearby mainland. Quoting from the report given to him by Quirós in Manila in 1596, Governor Morga wrote of the Marquesas islanders: '[T]heir boats, and their customs in other matters, do not suggest that they have come from very far away'.[9] Juan de Iturbe, who sailed with Quirós in 1605–06 expanded on this belief:

> seeing that those islands lay in the middle of that ocean, they [the Spanish] considered that they could not have been colonized from Peru nor from New Spain, which lay 600 leagues away, nor from the western part, in which direction they afterwards sailed for 1200 leagues. It seemed a very obvious fact and one which left no room for doubt that the mainland of the southern zone, whence these islands must have been colonized, was very near.[10]

If, however, the mainland was found not to be in the vicinity of the Marquesas, Quirós proposed to return to Santa Cruz and extend the search from there.

Quirós' conviction that the South Land existed was not

based only on these assumptions. Quirós was an experienced cartographer who would have been familiar with at least some of the world maps of his time, on several of which Terra Australis Incognita indisputably covered the earth's southern hemisphere. These undiscovered 'Austral Regions', he explained in a letter to the viceroy, lay between the Cape of Good Hope and the Strait of Magellan and extended to 90° South latitude, that is, to the South Pole. With a ship of 60 tonnes and a crew of 40 he could find this mainland. He was ready and should be sent on the voyage very quickly. This would be a discovery as great as that of Columbus.

Evidently Velasco did not respond, and in view of the events at Santa Cruz his hesitation is not surprising. Impatient and determined, Quirós wrote again, repeating his request and urging that he be sent 'without delay' to discover the South Land and establish a settlement. A third letter was yet more impatient. A curious sense of entitlement, an expectation of immediate compliance with his demands, seems to have taken hold of the previously restrained pilot. But he scaled down his requirements. If the cost of the ship he had requested was excessive, he would reduce the crew to twenty men. His motive was solely the honour of God and the service of his Majesty.[11] And his certainty of the existence of the vast antipodean continent was absolute.

The concept was, of course, magnificent—vast dominions to be added to the Spanish empire, millions of souls to be saved and brought into the Catholic Church. The viceroy of Peru, however, refused to assume responsibility for the expedition. He suggested that Quirós go to Spain to lay his plans before the king, and on 16 April 1598 himself wrote to King Philip II commending Quirós and his proposals. On the 17th, carrying this missive and letters of introduction to the king's councillors, Quirós sailed from Callao. A journey of 22 days up the South American coast brought him to the city of Panama, from where he crossed the isthmus to Puerto Bello on the Caribbean coast. Here he embarked on a frigate for Cartagena de las Indias, modern Columbia's Cartagena, from whence the great Spanish treasure fleets sailed for Spain. The city was in an uproar, for an English fleet of twenty large ships under the Earl of Cumberland had appeared before it, the force that had just taken the city of Puerto Rico and its massive fortress. With the appearance, however, of the Spanish West Indies flotilla, the English withdrew. A load of silver arrived from Puerto Bello and the fleet sailed for Havana.

Meantime, Quirós had despatched to Velasco a new, detailed plan for his proposed expedition, 'in case I should die'. Later he recorded, 'After twenty-seven days we anchored at Havanna, whence we sailed on the 16th of January in the following year, convoying thirty ships'.[12] At 29° North latitude a hurricane roared out of the Atlantic. 'Many ships disappeared, and others, including ourselves, were obliged to return to Cartagena on the 3rd of March.'[13] Quirós remained in Cartagena de las Indias for the rest of the year, waiting for passage.

On 4 January 1600 he joined a convoy of 27 vessels that sailed from Cartagena, carrying, in his words, 'fifteen millions' in silver. The Atlantic crossing was stormy, but off Cape St Vincent two English ships were captured and on 25 February, to the music of pipes, drums and coronets and the thunder of artillery salutes, the fleet came to anchor at San Lúcar de Barrameda on the Spanish south coast.

Quirós found that reaching King Philip III at the time was not possible, for the newly married young monarch and his queen were touring the country. There was, however, an attractive alternative. The year 1600 was a Jubilee Holy Year in Rome, and Quirós, who hoped to seek spiritual favours from the Pope for those participating in his expedition, decided to visit the ancient city. 'I sold the little I possessed, bought the dress of a pilgrim, and only with the help of a pilgrim's staff I went on foot to Cartagena of the east [on Spain's east coast].' An Italian galley brought him to a harbour town near Genoa. 'Thence, dressed as a pilgrim, and accompanied by two others and a friar, we passed through all the finest cities of Italy', finally reaching 'the great city of Rome'.[14]

Quirós now had the good fortune to meet a Dominican friar, Diego de Soria, whom he had known four years earlier in Manila. Soria spoke of Quirós to the Spanish ambassador to the Holy See, Don Antonio de Cardona y Cordova, Duke of Sesa, describing the pilot as an outstanding navigator with extensive experience in the Pacific. Sesa sent for Quirós and listened with interest to his account of the strange lands he had seen and his desire to return to them. 'I addressed myself chiefly to the importance of saving an infinity of souls, such as exist in that new world', Quirós recorded. Impressed, Sesa took Quirós into his household and during what became an almost 17-month visit, promoted his scheme. Of the pilot, Sesa wrote to the Spanish king:

I hold him to be a man of good judgement, experienced in his profession, hard-working, quiet and disinterested; his only excess is in being over-zealous for the service of your Majesty and for the public good which he hopes for from the voyage of discovery; I have at times seen him somewhat impatient at the delay, but this is a condition natural to those of his nation.[15]

The reference to 'his nation' evidently expressed Sesa's view of the Portuguese. In the case of Quirós the assessment was certainly accurate.

The duke arranged a meeting in his home of Rome's best pilots and mathematicians, who examined Quirós' records and charts, and concluded that the navigator's plans were feasible and should be adopted. Quirós believed that the Marquesas Islands lay between 9° and 10° South latitude, and the savants agreed that the Terra Australis mainland and any offshore islands were in 15° South. Sesa relayed to the Spanish king their affirmation of Quirós' extensive knowledge of navigation and his skill in cartography, even in making globes. Possibly at this time Quirós wrote a 'Treatise on Navigation', which the duke reportedly enclosed in a letter to the king.[16] In this exposition Quirós discussed the limitations under which a pilot worked, commenting on poorly made instruments and noting that two or more pilots would seldom agree on their observations of latitude, while longitude was simply an estimate arrived at through dead reckoning. All charts were defective, he said, and would be unless an experienced cartographer with accurate instruments continuously observed and recorded what he saw, a circumstance that would not come about for more than a century. He also described two navigational instruments of his own design, one apparently a type of compass, the other helpful in determining latitude. They had been praised by the Roman mathematicians and geographers who saw them.

Sesa obtained an interview for Quirós with the Pope, Clement VIII. Clement, then 64 years of age, was dedicated to the Counter-Reformation, which was directed mostly at the Protestant Reformation but also at reform and renewal within the Catholic Church. A major emphasis of the Counter-Reformation was missionary work in the regions of the world colonised by mainly Catholic nations, and Pope Clement was a man very much concerned with the spiritual functions of his position. He would have recognised with interest the fervent religious idealism of Quirós. Thus in the course of apparently several papal audiences, Quirós found an attentive listener to his plans to bring under Spanish colonial governance and into the Catholic fold the great undiscovered South Land and its untold numbers of heathen people.

Quirós received 'many graces and indulgences' for his proposed voyage, and letters from the Pope to the King of Spain. He was given rosaries and Agnus Dei medallions, little wax discs imprinted with the lamb symbolising Christ, that had received the papal blessing, and 'a piece of the wood of the Cross', about which there was a 'great difficulty' which he does not explain. In March 1602 Pope Clement wrote to the Franciscan prelates in Peru commanding them to send friars to evangelise the people of the Austral Regions 'recently discovered' by Quirós. With little doubt Clement understood that he had before him a true man of the Counter-Reformation, zealous, idealistic and fully prepared to devote himself to the propagation of his faith.

Quirós remained in Rome until April 1602, impatiently waiting for additional letters from Sesa to the princes and councillors of the Spanish court. It appears that during these months Quirós developed a mounting conviction that his return to the Pacific was a religious mission of immeasurable importance. The solemn pomp of the Holy Year, the mesmeric colour, sound, candlelight and gleam of precious and sacred objects in the great basilicas, the canonisation of Saint Raymond which he witnessed, the atmosphere at times of intense spiritual dedication in the crowds around him, probably especially his meetings with Pope Clement, would have struck deeply into his consciousness. In an already profoundly religious man, strong in his beliefs, these circumstances seem to have shifted the focus of his resolutions from territorial conquest for Spain to the glories of evangelisation and the conviction that he had been singled out by God to bring Catholicism to Terra Australis Incognita.

On the afternoon of 3 April 1602 Quirós left Rome, making his way north through the countryside of springtime Italy. His arrival in Genoa coincided with the sailing of a fleet of six galleys headed for Spain, on one of which he secured passage to Barcelona. Arriving in Madrid in the heat of summer, he learned that although the court had moved to the pleasanter climate of Valladolid, the king was in residence at the Escorial, the huge, austere monastery-palace built by his father, Philip II, 42 kilometres northwest of the capital city. There the letters he carried from the Pope and the Duke of Sesa gained Quirós entry. Received by the king, he was able to 'speak, and kiss his royal hands, and give him my memorial respecting my pretensions, on Monday, the 17th of June ... he heard me with his accustomed clemency and benignity, and replied that he would order the matter to be seen to'.[17]

Philip III was 24 years old, a pious and benevolent young man who tended to leave many of his responsibilities in the hands of his first minister and favourite, Francisco Gómez de

Sandoval y Rojas, later the Duke of Lerma. Clearly, however, the young king was greatly impressed by Quirós. The intensity of the navigator's belief in what he was proposing would have been arresting, and what he offered awe-inspiring—a vast new kingdom for the crown of Spain, untold millions of souls to be won to the Church and to salvation. Philip's grandfather, Charles I, had given his successors the great dominion of the Americas. Could this inspired pilot do similarly for Philip?

Whatever his thoughts, Philip was sufficiently intrigued to order that Spain's Council of the Indies, the normal channel for such endeavours, be bypassed and Quirós sent to various other officials, including members of the Council of State, an august body of experienced senior advisers, to whom Quirós gave his letters from Velasco and Sesa, his charts and other papers. Apparently there was an interview as well with the royal favourite, the future Duke of Lerma.

The response was mixed. Some glimpsed his vision and considered his scheme worthy of support; others thought, as he said, 'little of it or of me'. The usefulness of securing additional lands for an already extended empire and the cost of conquest and then of maintenance were questioned. Quirós commented later that he 'had much trouble at court', and resorted to submitting daily letters to the king, arguing for his enterprise. Obviously Philip remained interested and, never forgetting that he was king, made up his own mind.

On 31 March 1603 the young monarch signed a series of *cédulas* or royal orders which Quirós received on 5 April. Two *cédulas* to the president of the Casa de Contratación in Seville ordered that Quirós be provided with 'good passage' with the first available fleet sailing for the Americas, that he be supplied with duplicates of his own two navigation instruments, and that he be paid 500 ducats in an initial part-payment of 1500 ducats for his journey. A similar order to the viceroy of Peru regarding the 500 ducats is not explained, but documentation clearly confirms that a total of 1500 ducats was available to Quirós for his journey. In addition a letter to the viceroy, who may at this time have been the newly arrived Don Gaspar de Zuñiga y Azevedo, Count of Monterey, described the recommendations received on Quirós from the Duke of Sesa and the Pope, and ordered that the navigator be provided with two very good, well-equipped and well-provisioned ships 'without delay or raising difficulties', these to include crews, munitions and arms, goods for barter, and some Franciscan friars, all costs to be defrayed from royal revenues.

An additional *cédula* was addressed to the king's officers and officials throughout the empire, requiring that they protect and assist the men and ships of Quirós' voyage, should they arrive at any Spanish port. Yet another provided for the selection by the viceroy of Peru of an alternate leader of the expedition in the event of Quirós' incapacity or demise. A final order, dated 26 April 1603, entitled Quirós to take with him Esteban Borguñon, an expert in making navigational instruments, and a servant, Diego de Valle. No word, however, establishes whether or not these men actually accompanied him. For Quirós what had become the ultimate goal of his life—the spiritual conquest of the mysterious Austral Land—was in sight.

As Quirós' project became known, there was a curious echo from the past. Documents at Valladolid dated 1603 and 1604 record the request of the Chaplain of the Royal Chapel at Lima, Peru, one Sebastián Clemente, to join the expedition. His brother Francisco de Mondéjar had been a captain of infantry on the *Santa Isabel*, the *almiranta* of the Mendaña expedition of 1595, which disappeared. There were rumours of survivors, and the chaplain believed his brother to be alive on one of the Solomon Islands. No later document confirms Sebastián Clemente's sailing with Quirós in 1605, however.

The official business of the expedition having been settled, the Council members gave way to their evidently genuine curiosity, and the pilot was asked to explain his plans to them. The dignitaries assembled in a garden of the court and here Quirós spread his charts and expounded on his proposals. After this, as he later recorded, 'I set out on the road to Seville'.

Seville, the centre for Spain's overseas trade and exploration, was linked to the Atlantic Ocean by the Guadalquivir River, which empties into the Bay of Cadiz. Here the great galleons of the New Spain convoys rode at anchor in the San Lúcar de Barrameda roadstead. The fleet was ready to depart, but Quirós was delayed by bureaucratic procedures, possibly having applied at the Casa de Contratación for permission to take with him three female servants (*criadas*), which was denied.[18] Thus it was after dark before he caught a brigantine heading downriver, only to find at San Lúcar that the fleet had sailed. Fortuitously, he found a fast-sailing frigate that caught up with the convoy. This consisted of 30 ships, the flagship carrying among other functionaries a new viceroy for New Spain and his wife. Reaching the West Indian island of Guadeloupe after the Atlantic crossing, the ships anchored to allow their passengers to land on a deserted beach to hear mass. Quirós wrote:

> At dinner-time the persons of most consequence went on board again; but a great many remained on shore, wandering

about or washing clothes. They were suddenly attacked by the natives of that island, who fell upon them with great shouting and flights of arrows.[19]

There was a frantic rush to reach the boats and ships. More than 60 people drowned. There was further disaster. That night a strong southwest wind came up, driving against the ships still anchored close to the shore and to each other. The flagship and the galleon *Pandorga* collided violently, the viceroy and his wife, 'nearly naked', were hustled onto another ship and the two heavily damaged vessels set on fire, lest they fall 'into the hands of enemies'. The Caribbean was a popular hunting ground for foreign treasure-seekers.

Quirós' frigate now separated from the convoy and headed for the island of Curaçao, but struck rocks and shattered. All on board found refuge on the rocks and eventually reached Guayra, now La Guaira, in Venezuela, but it was eight months before Quirós obtained passage to Cartagena de las Indias. During this time he made a surprising discovery: the three orphaned children of a brother, Gaspar Fernández de Quirós, from whom he had not heard for many years. On his departure, Quirós took with him the two young boys, leaving the little girl with her grandfather. The boys, Lucas and Vicente, would accompany their uncle on his subsequent voyage. Of the girl there seems to be no further mention.

Despite bearing documents from the king, Quirós' journey to Peru was fraught with difficulty. He wrote that arriving penniless in Panama, he was sued for the unpaid rental of some mules, and injured when with a crowd of worshippers he attended the carrying of a Holy Sacrament to the upper storey of a hospital, where the floor collapsed, tumbling 60 people, along with beds and patients, some 20 feet to ground level. His injuries put Quirós in hospital for a cure that cost him 'four bleedings and two months and a-half in bed, without possessing a single maravedi'. Someone, unnamed, took 'pity on me in my necessity'.[20]

Finally embarking for Peru 'without a bit of bread or a jar of water',[21] he arrived in Paita from where he sent a letter to the viceroy. Embarking again, he arrived at the port of Callao on 6 March 1605. He was in debt for the passage and food but, hiring horses from an acquaintance, entered Lima that night. Unable to find a hostelry, he met a potter who allowed him to sleep among his pots for that night and 'three other nights'.

Quirós' much-mentioned poverty at this time needs to be considered. In a letter from the king to the viceroy of Peru dated 31 March 1603, Quirós is recorded to have received 1000 ducats prior to his departure from Spain.[21] An additional 500 ducats was available to him, and the total is a confirmed 1500 ducats. To determine the actual value of money in past centuries is very difficult, but a ducat was a gold or silver coin and it can reasonably be assumed that the amount was considered sufficient for the journey to Peru. Quirós' expenses during the long delays at Caracas and Panama cannot be estimated, and presumably the renting of mules and later hiring of horses were to provide transport not only for himself and his baggage but for his two nephews as well. Whether or not he was also accompanied by a servant or the instrument-maker Borguñon is not mentioned. It is therefore impossible to determine whether Quirós' insolvency was as complete as he claimed or possibly exaggerated. Whatever the situation, he did not mention the *cédulas* regarding money in his narrative.

Chapter Eleven
THE THIRD VOYAGE:
1605–1606

T hree days after his arrival in Lima, Quirós was admitted
briefly to the presence of the new viceroy, Gaspar de Zuñiga
y Azevedo, Count of Monterey. A formal audience was held on
25 March, attended by two judges, two Jesuits, the General of
Callao, the viceroy's captain of the guard, and a secretary. Quirós
read aloud from some of his documents and answered questions, the viceroy tracing the navigator's mention of places on
a map spread out on a 'buffet'. Apparently Monterey was satisfied, but being in ill health and with numerous other concerns,
he put the arrangements in the hands of others and, as Quirós
remarked, the 'despatch' did not go as quickly as 'necessary' or
'as I desired ... I had more opposers than helpers'.[1] Among
those who initially created difficulty was Fernando de Castro,
husband of Isabel Barreto, who considered the Solomon Islands

his inheritance from Mendaña. However, reasssured that Quirós had no interest in that archipelago, Castro became supportive.

The matter of a successor in command in the event of Quirós' incapacity or death was discussed. Orders from the king stated that the viceroy should designate a suitable person for this eventuality. Quirós objected strongly. 'I did not wish to take with me any one who would know he was to succeed me, for that was an arrangement fraught with obvious danger.'[2] He would, he said, in due time nominate his own successor. Apparently the expedition finally departed with the viceroy's sealed orders as to a successor, an issue that would arise in just over a year's time.

Questions of morality troubled Quirós. He wrote to the Archbishop of Lima, Don Toribio Alfonso Mogrovejo, on bringing back to Peru or to Spain some natives from the Austral Lands to teach them Spanish ways and the language, that they might serve as interpreters on succeeding voyages. Involved reasoning going back to the views of clerics of previous centuries finally approved the suggestion and dealt also with the ramifications of gift-giving, barter and baptism. Essentially, a fair and peaceful approach was called for. All dealings, however, were to be to the advantage of the Europeans, and their safety was to be assured at all times.

The organisation of the expedition was taking time. Maritime matters were in the hands of an admiral, Juan Colmanero de Andrada, who, Quirós felt, was 'not well disposed' towards him. He was not, he believed, receiving the immediate compliance with his requirements to which he was entitled. A sailing date in October was planned, but it became evident that this would not be met.

Quirós' sense of frustration mounted. He appears to have been driven by a conviction that God's hand guided his every move, by a belief in the absolute righteousness of his actions, however they might lack moderation or be in conflict with other issues. The restrained and forbearing chief pilot of Mendaña's voyage to Santa Cruz seems transformed into a man with an overwhelming sense of the importance of an achievement only he could bring about. In a letter apparently addressed in 1607–08 to the king, Juan de Iturbe, the expedition's accountant, wrote, 'if the said viceroy did not grant him immediately the supplies he requested, he would shout in the streets and squares saying that he carried orders for it from the Council of State and from the Pope',[3] the inference being that he, Quirós, was of greater importance than the viceroy, whose authority emanated merely from the Council of the Indies. This arrogant display caused the viceroy to form 'an unfavourable opinion' of the navigator, said Iturbe, and had not such considerable expenditures been made on the expedition, he would never have entrusted it to Quirós. Whether the viceroy would have diverted from his orders from the king remains a question.

Quirós later attributed the failure of the voyage to the delay in sailing from Peru, for which he blamed the viceroy. It appears, however, that on his side Monterey was concerned with Quirós' procrastination. In a letter to the navigator dated 21 November 1605, the viceroy noted that he had not received an anticipated progress report on the work for the expedition, and trusted that 'great haste' was being made towards its completion. He concluded, 'let me know whether the officials in whose charge it lies to provide all that is necessary for your equipment are doing so promptly, as I ordered them to do'.[4] That Quirós finally 'induced' the viceroy to appoint officials to assist him, as he wrote in his narrative of the voyage, does not ring true.

In fact, the viceroy was complying very generously with the stipulations of the *cédulas* he had received. Quirós was directed to examine the ships anchored at Callao and to make known his preference, and by 27 April had selected a privately owned ship named the *San Pedro*. A second vessel, also called *San Pedro*, was chosen, and by the end of June both had been purchased. The documents refer to them as *naos*, which at this time meant galleon-type vessels. The first, costing 13 000 *pesos corrientes*, became the flagship or *capitana*, and was renamed *San Pedro y San Pablo*. The second, purchased for 10 500 *pesos corrientes*,[5] was designated the *almiranta* or consort, retaining its name of *San Pedro*. The figures given for the size of these ships vary considerably. Diego de Prado y Tovar, who travelled with the expedition, stated them to be of 60 and 40 *toneladas* (tonnes burden) respectively. More likely are the figures given by Quirós at 150 tonnes for the *capitana* and 120 for the *almiranta*, although at a later date the Viceroy of New Spain quoted 120 tonnes for the *capitana*.[6]

Some reason for the discrepancies may lie in the fact that a *tonelada* was in fact a tun of wine, and a vessel's given tonnage was simply an estimate of the cubic capacity of the hold in terms of tuns of wine. At times this estimate may have been fairly rough, with an additional variable in the fact that the *tonelada* itself could differ in its cubic capacity, depending on where it was manufactured. Whatever the circumstances, however, these were small ships, vessels of 700 tonnes and more having been built in the viceroyalty since the 1590s, and Quirós made a point of saying that he had deliberately chosen the lesser ships. In overall length they probably would have been some 20 metres, and in width 6 or 7 metres. Both had three masts and bowsprits, the mizzen rigged with a lateen sail, the main and fore masts carrying topsails and mainsails with bonnets, and the bowsprit a crossyard and spritsail. Colourfully painted carvings of religious figures ornamented the stern of each ship. Both were most likely built in the dockyards at Santiago de Guayaquil in Ecuador, and after refitting at Callao were said to be the sturdiest and best armed in 'either sea'.

The viceroy ordered a further valuable addition to the little fleet: a government-owned 'English' *zabra* or launch, strongly made and reportedly a good sailer, that had come in from the Galápagos Islands with a shipwrecked crew. Smaller craft as part of an exploration group, with their usefulness obvious to any sailor, had been prescribed by King Philip II. According to Diego de Prado y Tovar, the purpose of this launch was to carry home news of the discovery of Terra Australis 'with all the speed due to His Majesty's service'.[7] It was named *Los Tres Reyes*, *The Three Kings*.

Six months were spent in refitting and equipping the three vessels. Provisions for one year included salted meat and fish, rice, biscuits, chickpeas and 'fruit and animals of Peru'. The *capitana* carried 800 jars of water, the *almiranta* 600. There were spare sails and additional sailcloth, probably the white cotton canvas woven in Peru, cordage, royal ensigns, hardware and tools, arquebuses, blunderbusses, and muskets with fork-rests, lead and iron shot, fuse-cord, powder in jars, bronze and iron culverins, lanterns and oil lamps, copper galley pots, pans and utensils, pitch, thousands of sailmakers' needles, hundreds of white wax candles, trade goods ranging from thousands of 'gewgaws of imitation gold and silver' to items of clothing, apparently of colourful materials, to dress native chiefs as gestures of friendship. There were also vestments and sacred ornaments for the mass, and notably, 'one copper apparatus, in two pieces, for extracting fresh water from sea water', a device invented by Quirós.[8] A payment of 4000 pesos to the apothecary Pedro Gómez Malo is recorded for medications packed in jars and baskets.[9]

As navigational instruments, Quirós had the same equipment as had Gallego on the first Mendaña voyage 38 years before— astrolabes and cross-staffs, compasses and sandglasses—with the addition of the devices he had designed himself. These, by order of the viceroy, were crafted for him in Lima. Very likely it was the compass of his own invention that was now placed in the binnacle of the *San Pedro y San Pablo*, but no direct mention of it is made in any further record of the voyage.

In the years between the two expeditions some improvements in navigation techniques had resulted from a better understanding of the errors in celestial navigation caused by the dip of the horizon and the sun's parallax. Efforts were being made in Europe to counteract these effects, but to what extent Quirós was familiar with them is not known. Nor is there any documentation of the charts he carried. He recorded having displayed maps to the pilots and mathematicians in Rome and to the viceroy in Lima, and it seems essential that he would have had representations of Terra Australis. Was he aware of the so-called Dieppe maps, depicting a large land mass lying south of today's Indonesian islands? We do not know. Perhaps of greatest importance was the advantage of having Gallego's records and his own experience of a prior voyage into the same region.

Only partial lists of the personnel who embarked on the three vessels have survived, and the figures given for their numbers vary. Quirós quoted a total of 300 people in one source and 158 in another, the latter figure confirmed by at least two other documents. Of these, 130 soldiers and sailors were on the king's payroll. Drawn from different parts of Spain's European empire, they included Spaniards, Portuguese and Flemings, almost invariably men of the lowest economic status, for whom there were few better options than the military or going to sea. One record of the distribution of 159 soldiers and sailors was made by Fray Martín de Munilla, Commissary of the Franciscans on board: 82 on the flagship, 61 on the *almiranta* and sixteen on the launch. In their loose shirts and wide breeches the seamen boarded with their meagre possessions in battered sea-chests, each man invariably with a knife in his belt and his own jug for the day's ration of wine.

The additional personnel, generally travelling at their own expense, would have included a number of gentlemen adventurers, or *entretenidos*, supernumeraries joining the expedition for the experience and hopefully for the wealth and glory of discovery. Usually they were members of the lesser nobility, who were available for positions suitable to their rank should such posts fall vacant during the journey. Until such an event they occupied no official capacity. Other sources make it clear that ships of the time also carried large numbers of 'grummets' or ships' boys, many of them young blacks, and page boys to serve the ranking officers. Such children were sometimes sent to sea by their impoverished parents; on occasion they were kidnapped, apparently at times on the beach of the Guadalquivir River in Seville. With Quirós' expedition there were also six Franciscan friars and, to care for the sick, four Brother Hospitallers of John of God. Mention is made as well of two privately owned slaves and on board the *San Pedro y San Pablo* two Chinese.

From the time of their hiring all crewmen were put to work

on fitting out the ships, and just prior to the expedition's departure each received a year's pay. It was generally possible for a man to mortgage this payment beforehand in order to make necessary purchases for the journey. In a final statement made on 13 May 1606 on the cost of the expedition, the accountant of the royal treasury of Peru recorded the total sum of 184 322 *ducados* (ducats) and 7 *reales* (11 *reales* to a *ducado*).[10]

It is during this period of preparation that there appears for the first time the name of the man who would be responsible for the expedition's most noteworthy achievement. On 9 July 1605, Quirós wrote to the viceroy requesting that one Luis Váes de Torres be appointed 'captain and pilot' of the second ship, the *San Pedro*.[11] Almost nothing is known of the background of this man. Prado y Tovar, who sailed with him, referred to him as a Breton, which at the time meant someone of Celtic blood, which in turn suggests that he came from the region of Galicia in Spain's far northwest. In this mountainous area, bordered on the west and north by the Atlantic Ocean, the Celtic tribe of the Gallaeci had lived for centuries. On the voyage Torres regularly led the armed shore parties, and comments that under his command the soldiers quickly assumed disciplined formations suggest that he may at some point have been a soldier. Quirós wanted 'none other' as captain of the *San Pedro*, and notably, 'the crew ask for him'.[12] In a letter to the viceroy Quirós asked that Torres be supplied with the rations and wine allotted to a commander, as at the time he was receiving those of a master. On the very day of sailing, Torres signed a document assigning to one Manuel de Barros power of attorney, authorising Barros to represent him in all current or future business dealings or lawsuits, to collect any moneys due to him and to free a

10-year-old mulatto slave boy. Of these arrangements we know nothing more.

Clearly Torres was a seaman of some reputation, respected and well liked, but again nothing is known of his experience. The choice was a fortunate one. His skill as a sailor and navigator and his steady competence as a leader would produce a significant increment to the geographical knowledge of the time, and specifically of what would eventually become the Australian continent.

On 20 December 1605 preparations were finished and the ships were ready to sail. Quirós and the captains of his two additional vessels, Luis Váez de Torres and Pedro Bernal Cermeño, travelled to Lima to bid farewell to the viceroy, who was ill. Quirós wrote:

> I asked him to pardon me for being so pressing, for it had been necessary to make a finish of my despatch. The Viceroy answered that, on the contrary, he was much pleased, and he embraced me and afterwards the other two captains.[13]

On the same day Quirós received officially the title of *cabo* (*cavo*) or commander of the expedition. Too sick to be in Callao for the departure, the viceroy provided Quirós with a letter to be read as his farewell. Monterey would, in fact, die a few weeks later.

On the following day there was celebrated the jubilee granted by the Pope, a special religious festival now held in the church of San Francisco in Callao, everyone dressed in the brown habits of the order of St Francis. Certain of Quirós' more elaborate plans had been curtailed by, he said, the 'envy' of others, but nevertheless 'all the people confessed and took the sacrament. The standards and banners were embarked, rolled up on their staves', and the six Franciscan friars issued from their convent to an emotional farewell from the crowd, many of whom, according to the pilot González de Leza, did not expect to see the voyagers again. Then 'we all went on board together . . . not a single man missing who had received pay'.[14]

At three in the afternoon Peru's admiral arrived and the viceroy's letter was read aloud. It reminded the explorers of the importance of their enterprise 'to the church of God, by the saving of many souls, and to the crown of Spain by the increase of its dominions'. All were charged to keep peace and good order, to obey their superiors and to regard their commander, Quirós, as if he were the viceroy himself. Any not doing so would be judged severely by the royal ministers and the viceroy. It concluded, 'May God guide you and send you forth to do His will'.[15]

Now came the chanting of men and the creaking of timbers as anchors were lifted, dripping, out of the water and loosened sails filled and strained. Amid gun salutes, flags and banners whipping in the wind, cheers and prayers from the shore, the three vessels turned their bows towards the Pacific. Quirós wrote, 'It was the day of St Thomas the Apostle, Wednesday . . . the 21st of December, 1605, the sun being in the last degree of Sagittarius'.[16]

Chapter Twelve

THE VOYAGE BEGUN

For seven days the ships ran without incident towards the southwest, steady southeast to easterly winds in their sails, making between twelve and 28 leagues a day as the men settled into their routines. The eight-hour watches were set. The galleys provided the day's hot meal each midmorning. The pilots took the angle of the sun, recording the ships' latitude and the estimated distance covered since their last observations. With Christian holidays following in rapid succession at that time of year, the journey began in a festive mood. 'The eves and days of Christmas, Circumcision, and Epiphany,' wrote Quirós, 'were celebrated with great festivity'.[1] The eve of the Epiphany, the 'eve of the name' of the little *Los Tres Reyes*, received special attention: 'many lights, rockets and fire-wheels' made bursts of brilliance across the darkening sea.[2] Leza, the chief pilot's mate, added that on the first of January there was also 'great festivity', apparently with a mock court and forfeits. In the happy enthusiasm of those first days of the journey, the voyagers would have looked forward with more certitude than ever to the golden promise of discovering the great South Land.

For the first three days Quirós was on deck or in his cabin making entries in his journal, but then he came down with an acute and totally debilitating headache. Quirós' recurring ill-health would weaken his leadership throughout the voyage. He did not, however, neglect his moral duty. On 8 January he had nailed to the mainmast a statement of the concessions granted to the explorers by the Pope, namely, a remission of sins even in dying without confession, so long as there was sincere contrition and the taking of the sacraments had been observed previously on certain specified holy days.

Some days later Quirós issued his instructions for the voyage, addressed to Torres as next in command, but promulgated throughout the squadron. Torres read that he was charged with maintaining 'Christian, political, and military discipline' aboard his ship, and that he was to approach the *capitana* each day to pay his respects. Taking the altitude of the sun daily and of the Southern Cross by night, to calculate latitude, the estimating of longitude insofar as was possible, and the recording of the ship's position on his charts, were routines Torres knew well. Throughout the fleet cursing and blasphemy were to be severely punished and playing with dice or cards, with the exception of backgammon, forbidden. This was not new to Torres. Gaming with dice or cards was, in fact, generally prohibited in Spanish ships, but he knew that the enforcement of such a rule was virtually impossible. That there was some incident aboard the *San Pedro y San Pablo* is suggested by the pilot Leza, who recorded that on 2 January 'it was ordered that we should throw the playing tables into the sea, which was done'.[3] Anyone taking God's name in vain forfeited his day's ration and if the offence was repeated would be fined or put in irons. A copy of the instructions was nailed to the mainmast. Those of the crew who could read would have done so for those who could not.

The day's ration of biscuit, pulses, drinking water, oil and vinegar, meat, fish or cheese was specified. At sunrise and sunset—and more often if necessary—two lookouts were to scan the horizon from the masthead, while at night watchmen were to be doubled, one stationed at the bowsprit. Each afternoon the ship's company was to kneel before the altar and led by one of the friars recite the Salve Regina and the Litany of Our Lady of Loreto, prayers for protection from the Virgin Mary that were traditional throughout the Spanish fleet. Those falling ill were to confess and make their wills. Signals made with flags and the firing of guns were outlined, and considerable detail described the handling of the natives of the lands they would encounter. Reflecting Quirós' past experience in the islands, the rules were essentially humane, but natives were to be handled with great caution, 'to be loved as sons and feared as deadly enemies'.[4] Importantly, enquiry was to be made as to whether they possessed gold, silver, pearls, spices or salt, whether the people of neighbouring islands were hostile or friendly and whether they ate human flesh.

Quirós' directives also specified the route his ships were to follow. They were to sail west-southwest to 30° South latitude, where Quirós believed they would find the great southern mainland from which, according to his theory, the people of the

Marquesas Islands had emigrated. If the mainland was not found, they were to criss-cross the Pacific westward between 10° 15' and 20° South. Failing to find land, the squadron was to follow the 10° 15' South parallel westward, in search of the island of Santa Cruz and the 'great and lofty volcano [Tinakula], standing alone in the sea'.[5] From Santa Cruz, where water, firewood and some foodstuffs would be found, the search for Terra Australis would continue. If the vessels were separated, the first ship there was to anchor in Graciosa Bay and await the other two. Should these ships not arrive, the captain was to raise a cross and bury a jar 'sealed with tar', containing an account of events and his intentions. He was then to continue his search to 20° South. Should no land be found, all ships were to steer northward and at 4° South follow that parallel west to rediscover New Guinea and from there continue to Manila and then through the East Indies to Spain, 'to give an account to His Majesty of all that has been discovered'.[6] There seem not to have been specific royal instructions as to course; apparently the itinerary was mainly Quirós' own. However, there were those in the expedition who argued for a more southerly limit for exploration. The account-ant Juan de Iturbe later wrote that Quirós had actually been charged with sailing to 35° or 40° South. Torres would later protest abandoning the search at under 30° South.

Who were the men occupying these three small vessels, headed across the almost unknown waters of the Pacific Ocean? For many the records preserve only a name and the man's duty. Nevertheless, as the events of the voyage unfold, some of them come across the centuries at least to some degree as individuals.

The master of the *capitana* was Manuel Noble, appointed by Quirós. Almost from the beginning of the voyage Noble suffered from excruciating headaches and was able to play little part in events. The chief pilot of the squadron and pilot of the *capitana* was Juan Ochoa de Bilbao, despite his surname said to be from Seville, and described by the accountant Iturbe as 'a fine sea man and diligent in his office'.[7] Evidently he had piloted the vessel on which the viceroy had travelled from Mexico to Peru, and Monterey had been impressed by the man's competence. On 29 May 1605 Quirós had written to ask the viceroy's permission to begin enlisting crew and proposed that Ochoa be appointed master of the *capitana* in order to handle the arrival of stores and provisions as soon as the ship was refitted. But the viceroy appointed Ochoa chief pilot, not master, and Quirós wrote: 'As Chief Pilot there came one against my will, whom they made me receive, as he had taken the Count of Monterrey from New Spain. He did me much injury'.[8] This was the first suggestion, without further explanation, of the enmity that would flare beween the two men.

A somewhat involved account written by Diego de Prado y Tovar states that Ochoa had been sentenced to six years in the galleys for unspecified misdeeds, a sentence commuted by the viceroy to service without pay on the voyage. As well, Ochoa was said to owe 14 000 dollars (16 000 dollars elsewhere in the record) in gambling debts to Lima merchants, and that Quirós had undertaken to deliver him at the journey's end to the authorities in Seville, who would then return him to his creditors in Lima. If this information is correct, it seems remarkable that the viceroy would appoint to the important post of chief pilot on a very long voyage a man under penal sentence and deeply in debt, unless for some unrecorded and important reason. This, however, seems not to have been part of the subsequent quarrel between Ochoa and Quirós. Other reasons for antagonism would come to the fore.

A shadow of estrangement also falls across Gaspar González de Leza, a Portuguese, who joined as the chief pilot's mate and second pilot of the *capitana* and who kept a surviving journal or *relación* of the voyage. He is later mentioned as involved with 'disturbances and commotions'. The search for Terra Australis Incognita would not be without dissension.

Quirós, however, had taken on board several of his own friends and supporters, a common practice among commanders. His nephew Lucas de Quirós travelled as a royal ensign, appearing on lists of those summoned to important meetings during the journey, while Lucas' brother Vicente is recorded as a sergeant. A cousin, Alonzo Alvarez de Castro, also accompanied Quirós, as did a close friend or possibly relation, Pedro López de Sojo, reportedly a grocer or an innkeeper but listed as an ensign, captain 'of the armed men' and sergeant-major.

Quirós also retained a secretary. This was the poet Luis Belmonte Bermúdez, about 21 years old, who had left Seville to seek his fortune in Mexico and later in Peru. Stirred by tales of the Spanish wars against the Araucanian Indians of central Chile, he wrote a panegyric on the gallant deeds there of the youthful García Hurtado de Mendoza, and carried with him on the voyage a copy of the epic poem *La Araucana*, by the young soldier-poet Alonso de Ercilla y Zúñiga. The poem describes vividly the fighting with the Araucanians and expresses warm sympathy for the bravery of the native warriors. The romantic disposition of young Bermúdez is obvious, and was no doubt reflected in his loyalty and devotion to Quirós, whom he apparently saw as a heroic visionary. It was Bermúdez who wrote

Quirós' journals, and the touch of the poet appears from time to time in an otherwise straightforward narrative of events.

On his return to South America from Spain, Quirós had evidently met in Puerto Bello Juan de Iturbe, whom a year later he proposed to the viceroy as overseer and accountant (*veedor y contador*) for the voyage. Iturbe received the appointment on 16 June 1605, and at the end of the voyage wrote his own *Summario Breve* or Brief Summary of events, a generally objective document although not without criticism of Quirós. The ship's surgeon was Alonso Sánchez de Aranda of Seville. The captain of the launch *Los Tres Reyes* was Pedro Bernal Cermeño, with two sergeants and a master gunner as his officers.

Of the gentlemen adventurers on board, Diego de Prado y Tovar is best known to history. Prado is first mentioned on 24 October 1605 as having received a grant of 500 *pesos corrientes* from the viceroy to enable him to participate in Quirós' voyage as a person 'necessary for the carrying out of the enterprise'.[9] He is referred to in this document as Captain Don Diego de Prado y Tovar and in his narrative of the voyage he claimed to have been captain of the *San Pedro y San Pablo*. This, however, is doubtful. The Franciscan commissary, Fray Martín de Munilla, refers to him simply as Don, as does Quirós in a later careful listing of each man's title and rank. Midway through the voyage Prado received permission from Quirós to transfer from the flagship to the *almiranta* under Torres, not the action of a ship's captain, although it has been suggested that possibly the viceroy meant him to be available as captain in the event of some failure on the part of Quirós. Quirós does not mention the transfer, but it is clear that Prado became part of the *almiranta*'s company.

Who, then, was Prado, who through his long and detailed

narrative would occupy a significant place in the record of the journey? Probably a man of noble background, he appears to have been a knight of the Order of Calatrava, Spain's oldest religious and military order, having hoisted his standard, 'white with a cross of Calatrava in the centre',[10] on the expedition's departure from Callao and making references to it later. Apparently he had some experience as a military engineer. When the *almiranta* later stopped at the Spanish-held island of Ternate, he is described as making improvements on the fortifications and doing so also in Manila. Prado had a predilection for enhancing the names of ships. In his *relación* he uses the nickname *San Pedrico* for the *San Pedro*, and the little *Los Tres Reyes* becomes *Los Tres Reyes Magos*, in effect *The Three Magi*. Prado seems to fit well the concept of a gentleman adventurer.

The number of such *entretenidos* on board the *San Pedro y San Pablo* is given in one source as four, but appears to have been six. It has been suggested that the men personally placed with the expedition by the viceroy—Ochoa, Prado and one Juan de la Peña Muñoz—were intended to curb what could be the excesses of the impassioned commander. There is no fair response to this other than to note that in Ochoa and Prado the expedition had acquired the seeds of conflict.

Luis Váes de Torres commanded the *San Pedro*, with the title of admiral. The master was Gaspar de Gaya and the pilot Juan Bernardo de Fuentidueña (Fontidueña), all men of experience and 'skill in seafaring'. The *almiranta* carried *entretenidos* as well. Don Juan de la Peña Muñoz was a protégé of the Count of Monterey, especially appointed by the viceroy to write a description of the lands discovered and a report to the king on events; his account, if it was written, has never been found. Don Alonso de Sotomayor was a second cousin of the Duke of Bejar, and evidently in middle age, as he claimed to have served the king for 24 years, fighting in the Indian wars in New Spain and subsequently in Chile. He was later transferred to the *capitana*. A third *entretenido* was Don Diego de Barantes y Maldonado, of whom little is known. Diego de Rivera was the ship's apothecary surgeon.

The Franciscan friars and the Brothers of St John were evidently divided between the two ships, with the elderly Commissary, Fray Martín de Munilla, chaplain and vicar of the fleet, on board the *capitana*. Munilla was described by his associates as of 'fervent zeal', having given up important positions within the order for a sea voyage scarcely suited to his advanced years. His writings, however, reflect a man of remarkable calm and objectivity. His journal is that of an attentive witness rather than a participant in the events of the journey. Interestingly, he mentions religious events only briefly. With the friars there was a young Peruvian named Francisco, serving as a *donado*, an assistant and sometimes interpreter, who wore the Franciscan habit but had not taken vows. Francisco was described as humble, devout and peaceful, and seemed to be exactly that.

Variable winds and a day of heavy seas forced temporary changes in headings, but essentially the ships maintained their course to the southwest and kept together, the lighted stern lanterns reassuring gleams through the night. By day gulls and boobies swept the sky, whales were seen and a man setting a sail aloft on the *San Pedro* lost his grip and fell into the sea. He surfaced and, grasping the rudder of the slow-moving galleon, climbed back on board. On 17 January the wind, blowing with force, swung to the southwest and west with a rising sea. Quirós decided to consult his pilots. He wrote:

the Captain [Quirós] presently showed a flag from the maintop
mast to take opinions, the weather not allowing of any other
way. The pilots of the ships said, by shouting, that, being outside
the tropics, all winds might be met with . . .[11]

The wind was indeed 'all around the compass', the ships strug-
gling to maintain their headings against massive swells from the
south. 'We made 15 leagues', the pilot Leza recorded on both the
18th and the 19th. They were, he calculated, 750 leagues from
Callao. At sunset on the 19th the boom of a cannon from the
San Pedro signalled the sighting of land to the south-southwest.
Darkness fell, but there was a moon, and Quirós ordered the
launch, with its lanterns lit, to proceed, the *capitana* and the *almiranta* following through the night. With lookouts watching
the gleam of moonlight on the dark, heaving waters, they
covered 20 leagues before sunrise flooded sky and sea. There
was no land to be seen.

By the 22nd the ships were in 26° South latitude, and there
occurred an event with far-reaching consequences for Quirós.
Critical as this was, it is not easy to understand because it was
recorded differently and in some records very briefly by the
men involved. Some circumstances are obvious. The wind had
swung to the south-southwest. There were rain squalls and
heavy seas. On the ships only foresails were set. For several days
dense banks of cloud had been seen on the southern horizon,
which to many seamen indicated land and in this case land of
considerable size. Fontidueña, pilot of the *almiranta*, and Ochoa,
the chief pilot, believed this to be the case. Prado wrote, 'I told
Captain Quirós . . . that it was a sign of great and lofty country,
but as it had not come out of his own head he did not take

much account of it'.[12] Nevertheless Quirós evidently called for a meeting of pilots, or in Prado's account all the ships' officers, but just when or even if this gathering took place is not clear. Prado describes it, but Munilla wrote that it was prevented by rough weather. Iturbe and Leza do not mention a council, nor does Quirós. Yet at some point Torres had boarded the *capitana* and discussed with Quirós whether in view of the weather the fleet should turn away from its planned goal of 30° South latitude. Apparently Torres believed that the fleet should continue to 35° or 40° South. Of his discussion with his commander, Torres later wrote: 'I gave a declaration under my hand that it was not a thing obvious that we ought to diminish our latitude, if the season would allow, till we got beyond 30 degrees. My opinion had no effect . . .'.[13]

Whatever the events of 22 January 1606, it appears that Quirós agreed that the *San Pedro y San Pablo* would steer southward for the cloud bank, the launch and the *almiranta* to follow. But at about midnight, Quirós appeared on deck to issue a startling change of orders. The helmsmen were told to swing the ship to a north-northwest heading. By Quirós' previous instructions, any change of course taken by the *capitana* would be announced by the firing of a cannon, with the other vessels then to follow. As far as is known there was no cannon shot. Conceivably the sound of the report could have been lost in the tumult of wind and waves. Or possibly in the wet conditions the gun could not be fired. Whatever the reason, in the thin grey light of daybreak the *almiranta* and the launch were seen far in the distance, pursuing their original course. Had the winds been different they could have been entirely out of sight. What happened next is not described, but obviously the vessels were brought together and sailed northwest, gradually decreasing their latitude.

Quirós apparently offered no real explanation. Later his narrative read, 'On the 22nd we were in latitude 26°'. Briefly he mentioned the 'timidity' of some of the crew. Then, 'We were obliged by the force of the winds and seas to stand on a W.N.W. course until we reached 25°'.[14] Were conditions actually making a southerly heading impossible? From the fact that the other vessels were continuing south, it does not seem so. Some crewmen may have been concerned, but most would have seen far worse weather. Torres' remarks on his earlier discussion with Quirós suggest that the commander had already considered 'diminishing' their latitude, that is, heading north despite his agreement to steer for the cloud bank. The midnight change of orders seems to support this. What then of the *almiranta* and the launch, which unknowingly continued south? Quirós' behaviour that night remains one of the enigmas of this complex man.

His action had serious repercussions, aboard the *capitana* especially. Since sailing from Callao there had been no landing. They had passed islets grown with coconut palms and rimmed with white sand, but obviously without the inhabitants whose presence would indicate fresh water and who might provide information as to other islands or a mainland. On 22 January the goal of the expedition had seemed to be in sight. For days the men had watched the horizon, increasingly convinced that beneath that long, dense cloud lay the promised Terra Australis—endless land, spices and precious metals, a welcoming population. Hourly they believed they were coming closer. Then in the darkness of night, at just 26° South the *capitana* had suddenly turned and now they were sailing away, the promise of the realisation of all their hopes sinking into the distance.

Incomprehension fuelled disappointment and anger. Mutinous talk among soldiers and crewmen ran through the ship and continued throughout the remainder of the voyage. There was anger among the officers as well. Iturbe, who was recording the events, castigated Quirós at length for failure to carry out 'so clear a duty' as to extend his search to 35° or 40° South latitude. There was no doubt, said Iturbe, that had they 'gone forward, they would have found in eight or ten days what they set out to seek'.[15] Prado appears to have confronted Quirós directly, accusing him of failure in the king's service, of being a man of 'little knowledge' who was giving an 'evil account' of himself. According to Prado, Quirós thenceforth 'took the greatest dislike and ill-will towards me', understandably enough.

Prado's sarcastic and confrontational comments, as well as the rancorous criticism subsequently expressed by others, would be extremely damaging to Quirós' later efforts to secure another expedition. Munilla wrote only that on the 23rd the *capitana* had altered course. Torres, in his brief report to the king, offered no further remarks on the matter. But no other discovery Quirós might make would compensate for this omission.

Three days later disappointment gave way to joy when in the afternoon land was sighted to the southwest, the lookout rewarded with a 'good prize', according to Leza. The ships were now seriously in need of water and firewood. Again the launch sailed ahead. The island was low, white sand backed by thick vegetation. No bottom was found at 300 fathoms. With darkness gathering, the ships moved away to a distance of three or four leagues, and through the night stood off and on with only foresails and mizzens. At daylight they found that the fleet had drifted to the leeward of a small and seemingly uninhabited islet. The island, their first real discovery of the voyage, was abandoned. Quirós named it Luna-Puesta at the time, but later changed the name to La Encarnación. Today it is Ducie Atoll, some 480 kilometres east of its better-known neighbour, Pitcairn Island. The Spaniards calculated themselves to be at 25° South latitude and 1000 leagues from Callao. With Ducie actually at 24° 40' South, their results with astrolabes and cross-staffs were reasonable.

Two days later a second island appeared. Green and flat-topped, it faced the sea with steep, sea-worn cliffs some 15 metres high. Leza wrote, 'The wind which blew over this island brought a smell of flowers and herbs: for they were abundant'. he ships moved in, 'so near the shore we could have thrown a stone on it, and no bottom at 300 fathoms'.[16] The launch approached the surf but was unable to come closer. Torres now lowered a skiff and sent it in with three men. Taking care not to lose their boat in the waves, they sprang onto the beach, and presently brought back 'certain fruits and herbs', among them the fruit of the pandanus, which the two Chinese crewmen recognised as edible and plentiful in China. There was no sign of human habitation. The ships circled the island and steered away to the west-northwest. Quirós called the islet Sin Puerto (Without a Port). Today it is Henderson Island.

For two days gale-force winds and pouring rain battered the fleet. Drenched and exhausted men struck the topmasts and struggled with sodden canvas while foaming water surged across the rolling decks. St Elmo's fire glowed at the top of the masts, watched with devotion by the seamen, who saw in it the saint guarding them. Fray Munilla, cross in hand, prayed and tossed into the sea on either side of the ship two Agnus Dei discs blessed by the Pope, perhaps those received by Quirós from Clement VIII.

In the wind-torn blackness of midnight there was a near collision with the launch, with much shouting and confusion. To avoid a similar occurrence, the *almiranta* fell behind and disappeared into the darkness. Alarm spread as bleak dawn light suffused the sky and the *San Pedro* was not to be seen. The *capitana* waited, rocking under bare poles, and by eight o'clock, to everyone's happy relief, the *San Pedro*'s masts and sails came up over the horizon. By noon she had rejoined the group.

Not having been able to take the sun's altitude reliably for three days, Quirós hoisted the flag that signalled a council. The vessels closed and Torres with his pilot Fontidueña, and Bernal, captain of the launch, came across the choppy water and climbed on board. Their estimates of the distance from Peru varied from 1110 to 1140 leagues. The fleet had sailed in an elongated curve some 5900 kilometres across the South Pacific, and it was agreed that if no mainland was soon encountered, a course of west-northwest would be followed in order to bring them to Santa Cruz, where they knew there was fresh water. Shortly afterward the masthead lookout cried out that land was in sight, 'the cause of great joy', in Leza's words.

Fear followed. An hour before nightfall a wild storm tore in from the east. In a darkness fiercely alive with lightning and thunder, the men had no idea of where the land was or how close they were to it. At dawn a dead calm descended. The Spaniards found themselves 4 leagues from an island, which was clearly without people or the possibility of anchoring.

Over the next several days the men sighted a series of uninhabited islets and atolls, 'low and level with the sea', inundated
'within the heart', and the sea around them 'without bottom'.[17]
Atolls are a geographic feature characteristic of tropical oceans
and were virtually unknown to the Spanish. Coral reefs that
commonly form around extinct, sunken volcanos, they typically enclose shallow lagoons, develop segments of low, flat land,
and are surround by great depths. Identifying them from the
descriptions of early mariners can be difficult, but those now
sighted by the explorers were the eastern atolls of the Tuamotu
Archipelago.

Fresh water had become a serious concern. On the *San
Pedro y San Pablo* twelve or fifteen jars of water were normally
consumed each day, and Quirós now reduced this to three or
four. He watched as it was poured, saw the hatch-cover locked,
and kept the keys, maintaining that fewer water jars than he had
ordered had been loaded.

His secretary Bermúdez wrote:

Presently he ordered a brick oven to be built over one of the
hearths, in order to make sweet water from sea water, with a
copper instrument he had with him. They got two or three jars
full every day, very good and wholesome.[18]

Leza also described the process:

6th. The hearth was arranged, and the apparatus for obtaining
fresh water from salt.
7th. The fire was lighted over the machine, and it began to give
fresh water with much ease . . .[19]

The ration was a *quartillo* (about half a litre) per man. Quirós served it out himself and the men accepted without protest. The next day was Ash Wednesday and Fray Munilla, in vestments, preached, blessed the water, and gave out ashes. Leza added, 'A fish was also grilled and divided amongst all.'[20]

A half litre of water each day for men living on salt meat and working under a blazing tropical sun was not enough. As well, word was circulating that the shortage had occurred because space intended for water jars had been taken by about 200 jars of white wine belonging to Quirós's friend, the innkeeper Sojo, who planned to sell it profitably in Manila. Aware of the growing anger, Quirós addressed the crew, according to Prado, 'in a doleful voice', promising them that the islands they had seen were signs of a mainland close by. He added:

> I give you my word that we shall be able to reckon ourselves the most fortunate men who have gone forth from Spain, for I will give you as much silver and gold as you can carry and such quantity of pearls as you shall measure them by hatfuls; for that of Peru and of New Spain is a very small matter compared with what I am telling you.[21]

Evidently the response was jeers and demands as to whether he had seen all these riches, to which he had to answer that he had not.

The next day the *San Pedro's* lookout sighted land and Torres had a signal gun fired. The explorers approached with few expectations, until from 3 leagues away rising smoke was seen. Ahead in the launch the men broke into shouts, 'People, people on the beach!' Coming closer, the Spaniards 'clearly saw men, and the sight was hailed as if they had been angels'.[22] Quirós was 'jubilant'.

The launch drew inshore, probing for a suitable bottom for the galleons. None was found, and the launch anchored, while boats from the ships continued the search. High surf barred the approach to the beach, where the islanders stood in a line, with clubs and lances in their hands. There was an exchange of signs, the natives seeming to encourage the Spaniards to land but warning them by gestures of rocks in the water and the breaking surf, a concern that puzzled the Europeans. Were they friendly or not?

In the *capitana*'s boat a young master gunner, Francisco Ponce, declared that they would achieve nothing if they avoided every danger; with his rosary around his neck he stripped, dived into the surf and swam toward the shore. On a rising wave he reached the shallows, where the natives, wading into the water, pulled him to safety. A seaman, Miguel de Morera, undressed and followed, and then two men from the *almiranta*'s boat. Their reception was astonishing. The warriors laid their weapons on the ground, and together bowed three times, and when a Spaniard was knocked down by the waves, picked him up, and embraced and kissed him on the cheeks, gestures of friendship that were found, Quirós remarked, 'also in France'. They then felt the Spaniards' bodies as if to assure themselves that the strangers were real.

The islanders themselves were naked and dark-skinned, with 'well-made limbs and good features'. They carried long wooden lances with tips hardened in fire, palm-wood clubs and heavy sticks. The Europeans' light skin, particularly that of one who was especially fair, fascinated them. A man who appeared to bc a chicf took a green palm leaf and gave it to one of the visitors. Some, wading into the water, seemed to urge the other Spaniards to come to their nearby houses, while women and children began crowding the beach. Eager to respond to this friendship, the Spaniards gave out what little they had in the boats—biscuits, half a cheese and some knives—and a few of their personal belongings. But the sun was lowering, and regretfully they returned to the boats. Several islanders followed them into the water, and the Spaniards invited them to get in. This they would not do, and the seamen rowed for the ships, climbing on board at sunset.

During the night the ships stood out to sea. Quirós, again bedridden, ordered that the vessels lie to so as to return to the same site the next day. Daybreak, however, found them 3 leagues to leeward of the beach with no chance of returning. Quirós held that the person responsible for this was the chief pilot, Ochoa.

The fleet was now lying off a long, narrow reef on which stood a small hill vegetated with coconut palms and other trees. Crews were despatched to look for water. Again the shoreline was rocky, with the breakers dashing heavily against the boulders. Holding their arquebuses high and laden with spades and crowbars, the men of the first boat jumped into water up to their waists just as the sea pulled back between surges. The last to jump was Quirós' secretary, Bermúdez, who was caught by the next incoming wave and plunged into deep water. The ensign and sergeant-major, Sojo, steadied himself with his hunting spear and pulled the young man to safely. To the admiration of his comrades, Bermúdez had not lost his arquebus.

The men of the second boat followed. Wet but pleased with

themselves, the group of some eighteen or twenty men formed up and marched to the grove of trees. Cutting their way with their swords, they emerged on the other side of the island to find 'a bay of still water', evidently the atoll's lagoon. Among the trees they found a number of brown stones, some stacked altar-like with plaited palm fronds hanging from the tree above. The men had come upon a *marae*, a sacred site, a temple and gathering place, but to them it appeared to be either a burial ground or a site where the Devil spoke to an innocent people. To purify the clearing they cut branches and set up a cross. They prayed and then dug without success for water, settling for coconuts to quench their thirst. Loaded with as many as they could carry, they set off for the boats. The island's main source of potable liquid, it seemed, was coconuts.

Among the trees there suddenly appeared an old woman, naked but for a loin cloth, blind in one eye and with the wrinkled face and body and few decayed teeth of someone apparently of great age. She carried some palm leaves, pieces of dried cuttlefish in a basket, a skein of thread and a knife fashioned from mother-of-pearl. A little speckled dog ran off. The woman offered no resistance to being taken in the boat and aboard the *capitana*. Quirós was delighted. Apparently unafraid, she ate meat and soup from a pot, but could not manage the hard biscuit until it was dipped in wine. A mirror startled but pleased her, and she looked with particular interest at the young boys and the goats. Quirós records that there was a gold ring set with an emerald on her finger, which she seemed willing to give him, but indicated that it could not be removed without cutting off the finger. She refused a ring of brass, but with a gift of white cloth and a hat, she was put ashore among the Spaniards.

The gold and emerald ring is not mentioned in the narratives of other members of the expedition. One can ask just how competent Quirós was in recognising an emerald or even whether the incident occurred in quite that way. Some modern writers, accepting Quirós' word, have taken the episode as evidence of a previous visit by Europeans.

Now five canoes carrying over 70 men approached the shore. Together with the old woman, the Spaniards came forward to meet them. Speaking rapidly, the woman recounted her adventure to the new arrivals, and obviously pleased, they disembarked without their weapons. With the Spaniards they walked to the sacred place, and here to the Europeans' great satisfaction, the islanders knelt before the cross when told to do so. Munilla wrote, 'As soon as they had seen us kiss it they did the same, for which we were most thankful to God'.[23]

Now the two parties returned to the boats and canoes on the beach. Several of the men, including the tall and impressive chief, seemed willing to go out to the ships. They got into the boats, but a short distance out panicked, jumped into the sea and swam away. However, when the chief attempted to do the same, the sailors restrained him. Coming alongside the *capitana* they tried to hoist him on board, to his absolute fury. Quirós descended to the boat, displayed a palm branch and removed the restraints. The chief looked sadly at the seamen and up at the masts of the ship, and pointed back to the land. Quirós dressed him in a silk shirt, breeches and a hat, gave him a tin medal and a case of knives and had him taken ashore. To the Spanish sergeant the chief entrusted a cap of hair and feathers to bring back to Quirós.

The two groups parted in their respective craft, the Spaniards saluting the others by firing off their arquebuses. Quirós named the atoll La Conversión de San Pablo. The identification of La Conversión has been disputed. One claim makes it the island of Anaa on the southwest edge of the Tuamotos; another identifies it as the more centrally located Hao.

The interaction of the natives with the Spaniards is difficult to explain. Ancestral ghosts and other spirit beings were part of Melanesian daily life, manifested in dreams, divination and magic. The suggestion has been made that at least some of the island people encountered by the Spanish took the strange-looking visitors for supernatural creatures, who arrived in vessels such as no man had ever built, and were therefore to be respected and propitiated. That the visitors were physically solid was often questioned until bodies had been prodded or shown to be skin and flesh under the clothing.

On their first approach to a shore the mariners were often signalled to land, the people on the beach showing little fear. On some occasions gifts of food were presented before they were actually requested. Undoubtedly, however, the presence of such beings, ghostly or otherwise, became alarming. Providing food for a large number of nonproductive visitors was without question burdensome, and the strangers' capacity for sudden deadly violence was destructive in the extreme. Agriculture and other aspects of life were disrupted when whole communities fled, even evacuating their islands. Not infrequently it was made clear that the Spaniards should leave. Elsewhere a chief drew a line in the sand which the explorers were not to cross. Solidly entrenched in their own convictions, the Spanish narrators of the various voyages speculated very little on the beliefs or social circumstances of the people they encountered, but did observe with interest the accoutrements of the islanders' daily lives. Some of the journals describe the manner in which houses were built, the type of weaponry, the finely woven matting and the cutting tools fashioned from shells. Understandably, watercraft were of special interest.

The island had been a happy interlude, if brief and without water, but Quirós' position remained difficult. He was again ill and in bed. The islands they now passed were small, uninhabited, waterless, but nevertheless he ordered that they be explored. This did not happen. Ochoa as chief pilot ordered them passed by. In the investigations into the voyage that took place afterwards in Mexico, the *entretenido* Francisco de Candategui stated that Ochoa had said stopping at the islands was pointless, as their search was for the 'mother of these islands', the mainland. Quirós was later to write angrily that he 'had seen that the Chief Pilot altered the course' he had ordered.[24] However, the chief pilot of a Spanish fleet did carry considerable authority. García de Palacio, author of the important *Instrucción Náutica* published in Mexico in 1587, advised that a captain should leave navigation wholly to the pilot unless he was incompetent.

Quirós' situation was increasingly serious. His friends warned him of incipient mutiny. Ochoa, he was told, was winning supporters by liberally giving out water and food. One rumour had it that Quirós was to be stabbed and thrown overboard. Arguments among the officers and clandestine meetings by day and at night intensified his suspicions.

There was another element to some of the opposition to Quirós: his Portuguese nationality. Spain and Portugal were separate kingdoms under one king, Philip III, their citizens to be regarded as equals. But Spain had been the larger, stronger nation, which had conquered its Iberian neighbour less than 20 years before, and there remained too some of the antagonisms and jealousies of past centuries. Prado, especially, made no effort to conceal his disdain of a Portuguese captain whose seagoing experience had begun on merchant vessels, not in the king's service.

It appears that Quirós was persuaded to take some action. He had two new orders fixed to the mainmast:

> No person shall take the name of God in vain, on pain of a fine of one dollar, for the souls.
>
> No person shall dare to put his hand to a knife or any other weapon whatever, on pain of a fine of thirty dollars.[25]

He mustered the crew and reminded them of the expense put forth by the king to send out the expedition to find the 'unknown part of the earth', and search for it they must, 'ploughing the ocean with long turns until it was found, even if it cost all their lives'.[26] To Ochoa he spoke at length of a chief pilot's duties. Through Fray Munilla, Ochoa then asked to be transferred to the *almiranta*. Quirós told him to go, but Ochoa changed his mind and remained aboard the *capitana*.

One night a commotion brought Quirós from his cabin to find several men fighting, among them Ochoa, who had wounded a man with his sword. So weak that he could scarcely speak, Quirós made no effort to discover cause or guilt. His recurring incapacity, his abandoning the southward heading at 26° South and the seemingly endless waterless islands they encountered were already an unfortunate combination of circumstances for the commander. Now he had failed to take charge in the face of violence.

The fleet was now on a northwesterly heading, seeking the island of San Bernardo, probably modern Pukapuka, which Quirós had sighted on his 1595 voyage with Mendaña. Their progress was slow. Fearful of unknown reefs, the ships lay to through the night and made sail at daybreak. Criticism was rife. They were sailing a course parallel to the coast of the mainland they sought, and would therefore never reach land. Or they would perish in terrible winds. The men's lack of dedication distressed Quirós. He wrote: 'these sayings were witnesses of the little love some had for the service, and of the great love they had for themselves . . . they were far from having valorous minds'.[27] Brave, devout and committed, Quirós failed to measure other men with realism and lacked the dynamism to inspire them to rise above their mundane desires.

In the afternoon of 21 February 1606 an atoll was sighted which Quirós declared to be San Bernardo. Leza took the sun at 10° 30' South latitude. The islet's position is actually 10° 30' South, 150° 15' West. At sunrise the next day the launch began the search for a suitable anchorage for the galleons. Failing in this endeavour, the crew joined landing parties from the ships digging for water among the palms. Only brine oozed into their holes. Coconuts were plentiful, as were birds and fish, which the men killed with swords and sticks. Torres, Prado and Munilla did not believe the islet to be San Bernardo, but a new discovery, and it is in fact modern Caroline Atoll, part of Kiribati. Despite the presence of the remnants of an old canoe, it was uninhabited, and was still deserted when rediscovered in 1795. More recent examinations, however, have uncovered basalt adzes and the ruins of a *marae*, so at some time it was occupied. The northwest heading was resumed.

On the night of 1 March 1606, the men on the launch fired two falconets and displayed two burning torches—the dark outline of land had been seen. A point of firelight appeared ahead, seemingly in response, and shortly, as the moon rose, land became visible not more than a league away. At daybreak an extraordinary reception commenced. Two canoes approached the *capitana* briefly, apparently to reconnoitre. Then ten small canoes came swiftly out from the shore, converging on the flagship, the men 'singing to the sound of their paddles, one of them leading, to whom the rest replied'. Alongside the ship, some stood up, shouting, dancing and gesturing, in what seemed to the delighted Spaniards to be the 'merriment' of an astonishingly friendly welcome. They were handsome people, well-built, tawny-skinned, some with long, bleached hair, one young boy described by Bermúdez as 'very beautiful', had he been a girl.[28]

The launch, meantime, had anchored close to shore, where a crowd of several hundred men and women had emerged from a village to line the beach. Suddenly a group of men began an extraordinary effort to seize the launch. With great speed some fastened a line to the bowsprit while others dived to secure ropes to the anchor cable and men on the beach began pulling. The startled Spaniards cut the ropes, which the natives rapidly rejoined. Several times the ropes were cut and rejoined, until the Spaniards fired their arquebuses 'without ball'.

The shocked natives scattered, but realising that the noise had injured no one, they returned, shouting and dancing, now carrying clubs and 6 metre lances and clamouring for 'all they saw, down to the arquebuses and swords'.[29] The *almiranta*'s boat, attempting to chase off the swimmers, had no effect. Leza wrote, 'until this time we had done them no harm',[30] but now the

Spaniards opened fire 'with ball'. Several natives fell and were carried to the village. The conflict ceased, the beach emptied and the seamen resecured their anchor lines. At the ships, lying farther out, an old man in a canoe had launched his own attack on the *San Pedro y San Pablo*, brandishing and thrusting with a very long and thick wooden lance.

> He made fierce faces with his eyes and mouth. In a very loud voice he seemed to order us to surrender . . . two muskets were fired off. The others cried out and threw up their arms, but he made light of it. With great pride he showed more signs of his anger.

Finally he stopped and joined the canoes still milling about the ships' boats. Here swimmers were grabbing at the oars, trying to force them from the sailors, and 'such was the courage and audacity of the old man' that he joined the fray by attacking an ensign, who took the blow on his shield.[31]

In the meantime Quirós and Torres had conferred aboard the *capitana*. Several journals make it clear that the two captains decided that if fresh water could be had, they would discontinue the search for Santa Cruz and instead steer south and south-west, 'traversing unknown parts in a higher latitude' in the quest of Terra Australis. But the immediate problem was finding a source of good water among fiercely hostile people. A well-armed party would have to be sent ashore, and apparently the men asked Torres to lead them.

The following morning the ships' boats, each with 25 men carrying shields and arquebuses, pushed off from the galleons, and escorted by the launch rowed for shore. On the beach the native warriors waited in orderly formation, holding their long lances, among them a few women armed with staves. Seeing the Spaniards ready to land, they brandished their weapons and shouted 'as if they had been drilled' a war cry that was, in Bermúdez' words, 'very brief but terrible'.

Torres shifted his party to a slightly more advantageous position, had the launch remain within gunshot range, and ordered his men to leap into the surging, waist-deep water, he being the first. One boat capsized with four rowers underneath but was righted by another wave so the men surfaced, although water jars, tools and guns were lost. Now the islanders attacked. Getting his men onto a reef where the water was less deep, Torres had them rapidly form up and discharge a volley. One or two attackers were killed. The others ran toward their village and, splashing their way onto the beach, the Spaniards followed. The houses seemed deserted, as the warriors, taking to canoes pulled up along the island's central lagoon, crossed to the opposite shore.

Several elderly men then appeared bearing burning torches and green boughs. Torres received them warmly, had their leader dressed in silk and asked him by signs to lead them to fresh water. This turned out to be a very small spring at a considerable distance from the shore, inconvenient and inadequate for the needs of the expedition. Here there was another skirmish, a Spaniard being wounded and the natives put to flight.

Torres took no captives; it could not have been done without further fighting, although one young boy, in Bermúdez's words as beautiful as 'a painted angel', seemed to offer himself as a captive or servant, and briefly Torres considered taking him on board and making him a Christian. There were other unexpected encounters—a nearly blind old man helped from his hiding place by the ensign Gallardo, a 'lady, graceful and sprightly' who did not prove to be 'prudish'. But despite efforts to maintain peace, there were several clashes, aggressive soldiers confronting angry natives, so that at least two warriors were killed and others injured.

Nothing constructive could be achieved, and Torres ordered a retreat to the boats. Again there was the struggle through the surf. A boat was swamped and the small supply of water lost. Some men were badly bruised and others, having trodden upon sea urchins, were in agonising pain. The ships were waiting at a considerable distance, and it was two o'clock in the morning before they boarded, Torres going straight to his *almiranta*. Quirós, he knew, would be annoyed that no islanders had been taken. Bluntly, Torres named the island La Matanza, The Killing.

Nonetheless, Torres had been much impressed by the island and its people. He had seen neatly cultivated plots, houses though small with lofts where the family slept. Floors were covered with mats finely woven from palm fronds, while matting of an even more delicate texture was used for the men's loin cloths and covering garments for the women. Cutting tools, fish hooks and other implements were made from the only available material, mother-of-pearl shells. Large double canoes were constructed almost entirely from materials derived from palm trees. The people were striking, light-skinned, the men tall and robust, the women, according to some of the Spaniards, as beautiful as those in Spain. Quirós named the island Peregrina or Pilgrim, but his frequent references to its handsome people led some later chroniclers to call it Gente Hermosa (Beautiful People). It is, in fact, Rakahanga Atoll, lying at 100° South latitude and 161° 06' West longitude in the northern Cook Islands.

The fleet continued its westward search for Santa Cruz, for 34 days maintaining an almost steady course on the 10° 10' or 10° 20' South parallel. The mainly easterly winds were light and many days the vessels barely moved on a flat, metallic-looking sea. Leza wrote:

> We made 5 leagues to the west . . . We made about 8 leagues . . . We suffered much from want of water, and the machine could not produce any for want of fuel, for we had come to an end of it and none for cooking. Occasionally, when some rain-squall came, it was received with great applause, the ship being spread over with sheets to catch the water . . .[32]

There were signs of land—birds, floating fruit, a long pole, pumice which they thought must come from the volcano near Santa Cruz, and snakes the Spaniards assumed were land crea-tures—but the horizon remained an empty blue line. Leza continued:

> At this time we must have gone, by imagination and dead reck-oning, 2120 leagues, seeing that we have not arrived at Santa Cruz, it would seem that we have given more distance to the ships than that which they had actually gone over . . .
>
> Or, perhaps, when they discovered Santa Cruz, those who discovered it may have thought that they had not gone so far and put it down E. of the true position . . .[33]

On 22 March they watched a total eclipse of the moon.

On board the *San Pedro y San Pablo* fear and anger focused upon the commander. He was accused of having deceived the

Pope and the king and of now leading his expedition to die
in these 'great gulfs' of the sea. Torres told Quirós to put the
mutinous individuals on trial, but the commander seems to
have accepted the situation with resignation. The fact was that
virtually every commander of a long-distance voyage of the
time had difficulties with his crew.

On 25 March, however, the smouldering resentment came
to a head. Ochoa announced that he had calculated the distance
from Callao to be 2220 leagues and with Santa Cruz, reputedly
at 1850 leagues from Lima, not yet sighted, the situation was
dire. Angered, Quirós signalled for a council. The other vessels
closed and Torres, Fontidueña and Bernal came across in their
boats. The men compared their charts and notes. As these were
compiled by dead reckoning, again there were discrepancies.
Torres made it 2000 leagues from Peru, acknowledging that
unknown currents could have slowed the ships, Santa Cruz
could be farther from Lima than had been charted, or he himself
was in error. Ochoa pointed out that they had been sailing for
94 days; Santa Cruz had been reached on the previous voyage in
69 days. Quirós declared that there was an error of 600 leagues
in Ochoa's calculations, commenting that Ochoa's past experience at sea was limited to coastal sailing. Apparently there was
a mistake in Ochoa's computations, but Quirós seems to have
exaggerated the figure.

Recriminations mounted, Quirós increasingly convinced
that Ochoa was one of the would-be mutineers. Torres, who
agreed that the *capitana*'s crew were mutinous, urged Quirós to
try and if necessary punish the pilot and any others involved, but
the commander refused. The timing of what followed varies in
the accounts, but at some point Quirós had Ochoa transferred

to the *almiranta* and, according to Prado, ordered Torres to have the pilot 'garrotted at once and cast into the sea after confession'.[34] Ochoa was transferred, but without a written order to do so, Torres refused to kill him.

The fleet now needed a new chief pilot. Separate narratives give different versions of events involving Pedro Bernal Cermeño, captain of the launch, and Gaspar Gonzalez de Leza, chief pilot's mate. It appears that Cermeño was briefly chief pilot, and then received the title of admiral. Leza then became chief pilot. Reinforcing his hard stance, Quirós had an execution block rigged on the yardarm. He regretted, he said, not having brought 'irons, fetters, and chains from Lima' and now 'lived with the caution necessary among such villains'.[35] Almost hourly he was at the binnacle, checking the needle's position. A request from the crew to allow some gambling was turned down. There were books on board, Quirós said, and 'one who would teach them to read', while practised soldiers were available to train the inexperienced in handling swords.

The journey west continued towards an empty horizon. Rain relieved the immediate need for water, one heavy downpour filling 50 or, by another account, 70 jars, but the expedition remained without firewood. Then at three o'clock in the afternoon of 6 April there was the excited cry of 'Land! Land!' from the *capitana*'s lookout.

The ships lay-to through the night, but at dawn, land, 'high and large', was seen at a distance of about 8 leagues, and the little fleet moved closer. When rising smoke was sighted, the explorers' joy was 'incredible'. People meant water.

With the launch and the ships' boats, and with 50 arquebusiers and shield bearers, Torres spent the greater part of the next two days reconnoitring the shoreline. They entered a lagoon, where they found a village enclosed in a wall of hand-laid stones, standing fortress-like on a reef. It seemed deserted, but as the Spaniards approached in their boats some 150 warriors armed with bows and arrows confronted them. An arquebus was fired into the air, and the natives fled, some diving into the sea. One man remained and showing no fear waded out to the boats, 'the water up to his neck'. Torres received him courteously. Gesturing with downward motions of his hand and calling loudly 'Pu! Pu! Yac!', the native made it clear he wanted the guns lowered, whereupon the Spaniards signed that the bows and arrows should also be laid down. The returning warriors did so, and at Torres' request, the man ordered them to scatter. Alone with the Spaniards, he signed that he was Tumai (Tomai), chief of the island, and 'in his own way made peace, interlocking a finger' with Torres.[36]

The Spaniards landed and Torres had the chief dressed in silk, which seemed to please him. In a further expression of amity, Torres and Tumai exchanged names. Torres then asked for water and promised that the village, which was given over to the Spaniards, would not be set on fire. He and his men, he told Tumai, would leave in five days. At the same time he posted guards, weapons found in the houses were stacked at the guard post in view of their owners, to be returned when the Spaniards left. Then he sent the boat to report to Quirós.

The behaviour of the islanders had made it obvious that they understood the danger of the arquebus, which they called *pu*. As soon as the ships had been sighted, they had rushed women, children, the elderly and their household goods into the inland woods. The Spaniards soon realised that these people knew of events at Santa Cruz during Mendaña's attempted colonisation eleven years before. According to Prado, Tumai himself had been there at the time, but other Spanish accounts do not mention this.

The island on which the Spaniards had landed was Taumaco, largest of the Duff Islands, a small chain of nine volcanic islets some 133 kilometres to the northeast of Santa Cruz. The voyage from Callao to Taumaco had taken three and a half months. Fourteen islands, mostly atolls, had been seen, none yielding fresh water.[37]

With the news of a peaceful reception, the ships were moved closer to shore and the launch anchored nearer still. It was Easter Sunday, and all the Franciscans landed. An altar with a rich canopy and handsome altar pieces was erected in one of the houses, and mass was said for six days in succession so the sacrament could be taken by almost all the explorers. To the Spaniards' extreme pleasure, the islanders who were present attentively imitated the kneeling and rising of the Christians. Quirós grieved that a people who seemed so ready to accept the Faith would be denied it.

Meantime, the watering of the ships was underway, boats and canoes plying steadily between the ships and the beach with the water jars. At Tumai's request he was brought on board the *capitana*. Quirós received him well, later describing him as a man of about 50, with a 'good body and face . . . beard and hair turning grey. He was grave and sedate . . . what he promised he performed'.[38]

They sat in the gallery at the stern of the flagship, with a table prepared 'that he might eat', which, however, he declined to do. Munilla was presented to him, Quirós kissing the Commissary's hand and having Tumai do the same. With gestures, Quirós elicited answers to a number of questions. Tumai said he had not seen ships or people like the Spaniards before, but had heard of them. The volcano near Santa Cruz island was just five days' sail from Taumaco. Quirós asked about other lands, and Tumai's knowledge of his region was obviously extensive. Pointing and counting on his fingers, he indicated the direction from Taumaco of some 60 islands, while Quirós, with a compass before him, wrote down the bearing from Taumaco of each. Tumai explained the size of the islands with big and small circles, throwing wide his arms for a very large land. Their distance he explained by acting out the number of nights it would be necessary to sleep to reach them. He described the colour of the people, and 'gave it to be understood that in one island they ate human flesh, by biting his arm, and indicated that he did not like such people'.[39] Taumaco was at war with many of the islands he named.

The next day Quirós went ashore with paper and his compass, and assembling a number of people, asked further questions, to confirm the chief's information and to add to it. The sight of someone reading astonished them, and the paper was carefully examined on both sides.

Mutual confidence was such that a number of soldiers and seamen visited the villages alone without causing offence, washed their clothes in the streams undisturbed, and any objects they left behind, such as clothes or pots or kettles being scrubbed, remained untouched. Torres commented that 'we agreed very well'. He wrote of the people as 'very great seafarers, all very bearded, very great archers and hurlers of darts; their very large boats could go a great way'.[40] Nevertheless, Tumai obviously believed in the possibility of violence against the visitors from his own people. He almost never left the Spaniards, and when he did so, placed with them someone else as, in effect, a hostage.

On the fifth day Tumai asked Torres to observe his own stated limit of five days on the island. Torres explained the need for firewood, and the chief summoned people 'with mighty shouts'. Shortly canoes were carrying out to the ships all the wood they could conceivably need. Torres also asked for coconuts and they brought over 300, not counting those already obtained through barter. On the seventh day the Spaniards began to embark. Quirós ordered four natives to be taken on board, the chief inadvertently witnessing this final sad and disillusioning act.

An interesting incident among the Spaniards took place on Taumaco. Diego de Prado y Tovar transferred from the *San Pedro y San Pablo* to her consort, the *San Pedro,* under Torres' command. The explanation occurs in Prado's *relación*. He wrote that he had become aware of the plans for mutiny aboard the *capitana*, and knew the identity of the leaders. He told this in confession to Fray Munilla, who said he knew of the situation and had passed the information to Quirós who, however, ignored it. Prado knew, he said, that the would-be mutineers wanted him as their leader, but not wishing 'to mix in such conflicts and lose the honour he had gained in the service of His Majesty',[41] he asked Quirós' permission to take himself, his rations and his belongings to the *almiranta*. Quirós acceded, in Prado's words, 'to get rid of the bother'.

Whether Prado knew as much as he claimed about a possible mutiny remains a question, as does the extent to which he himself may have encouraged dissension. There was the fact of Prado's thorough dislike of and undisguised disdain for his commander. A self-centred man, Prado may simply have tired of Quirós, but provided a more colourful reason for his departure. There can, however, be little doubt that the commander considered himself well rid of an arrogant and insolent passenger. The veteran soldier Alonso de Sotomayor left the *San Pedro* to join the *capitana* in Prado's place. Curiously, the next day the *capitana*'s surgeon, Alonso Sanchez, also transferred to the *almiranta*.

At ten o'clock on the night of 18 April the fleet set sail. Within two days, while still off Taumaco and a neighbouring island, three of the captive natives had jumped overboard and swum for shore.

SPE ET METV.

Chapter Thirteen
LA AUSTRIALIA DEL ESPÍRITU SANTO

1589

For four days the ships steered southeast and south, in the direction in which Tumai had indicated there were islands. They passed the wooded island of Tucopia (Tikopia), where Torres, approaching in the launch, was met by two men in a small canoe who presented him with a long strip of bark material 'like a very fine handkerchief'. On 24 April in a north-westerly breeze they altered their course to the southwest. Quirós records that asked what heading should be taken, he had answered, 'Put the ships' heads where they like, for God will guide them as may be right.'[1] Presumably Leza as pilot exercised some discretion in relaying this order to the helmsman. A day later they entered a region of high mountainous islands beyond which there appeared to be extensive land. The sense of excitement felt by the men pervades their narratives—Leza describes a chain of mountains reaching as far 'as the eyes could descry'; Munilla, 'land, which seemed to be very great'; Quirós, 'land of many mountains and plains'.

Columns of smoke rose into a silky blue sky and fires gleamed through the night. Moving closer, they could see cultivated fields and groves of palms and plantains. Men in canoes approached the *San Pedro* and exchanged fruits and coconuts for 'a good return'. Repeatedly the islanders called and gestured for the Spaniards to land. The voyagers were entranced.

Off the island the Spaniards called La Virgen María (Gaua or Santa Maria on modern charts), twenty soldiers and their officer, Sojo, followed the coast in the *capitana*'s boat and saw 'copious and broad rivers' flowing into the sea from 'beautiful ravines'. On the beaches groups of inhabitants called to the strangers to come ashore.

Munilla wrote:

> When they saw that we were unwilling to do so, a *cacique*, one of the head men of the island, who had been gazing at us from the top of a high hill . . . with unbelievable determination came down to the beach, jumped into the sea alone and swam to the *capitana*'s boat. He got in and spoke in an extremely arrogant manner.

Obviously he was not understood, and turned to go, whereupon 'our men bound him and brought him to the *capitana*'. Munilla added, 'The spirit and arrogance with which he spoke were a sight to behold'.[2] Impressive too were the bracelets of white boar's tusks around his wrists.

On board the ship he was given meat and biscuit, which

he ate readily. Quirós questioned him on the surrounding area, and he pointed to several places on the horizon, counting on his fingers. Strangely, according to Quirós, he concluded with the words 'Martín Cortal'. Martín Lope Cortal had been a pilot on Miguel López de Legazpi's ship sailing from Mexico to the Philippines in 1564–65. On a subsequent return trip to Mexico, Cortal and others were reportedly marooned on islands called the Barbudos.³ Quirós offered no further comment and no mention of this is made in any other journal of the voyage.

To keep his captive from jumping overboard Quirós had him put in the stocks for the night, intending to return him to land the next day, dressed and with gifts in order to spread word of the goodwill of the newcomers. At about nine o'clock that evening, the launch arrived alongside to report that another man, similarly arrogant in manner, had climbed aboard the launch, whereupon the soldiers had fastened a chain to his leg. With prodigious strength the captive broke the chain with his hands and, with the padlock and some links still attached to his leg, had leaped into the sea about two leagues from land. In the darkness, the boat's crew had not pursued him and returned to the ship. About eleven o'clock, the sentry at the *capitana*'s bowsprit heard despairing shouts from the water. Following the cries, the seamen found the man and brought him on board. The chain and padlock were struck off, he was wrapped up warmly and given food and wine, and then put in the stocks with his countryman, where they reportedly slept through the night.

In the morning Quirós made a show of reprimanding the man who had placed them in the stocks, implying he had known nothing of this indignity. He had the two prisoners shaved and barbered, their finger and toe nails cut with scissors—an instrument which fascinated both men—dressed colourfully and sent ashore. In the boat they both sang, and on the beach were joyfully received by some 300 men, women and children. Among them were the chief's two wives and young child, whom the Spaniards were allowed to embrace and kiss, which 'caused such rejoicing all around that it was worth seeing'.[4] Here again arises the question of the natives' concept of who or what the Spaniards were. Were they seen as spirit beings whose attention to the child was a good omen? Or was it simply a wave of joyous relief in seeing this demonstration of warmth in dangerous strangers? Fruit and a pig were brought to the Spaniards, whether as rewards or conciliatory offerings is difficult to say.

Encounters with other groups of natives were less amicable. Some, calling for the boats to come ashore, became angry when they did not, and fired off arrows believed by the Spaniards to be dipped in poison. The boatswain's mate, Francisco Machado, was hit in the face, but the arrow was spent and stopped by the cheek bone. The point was pulled out and the wound, inspected for poison, was clean.

On 30 April the fleet crossed open water, Leza taking the sun at 15° 10' South. To the southeast rose what appeared to be a massive mountain chain, with summits lost in clouds. Leza wrote, 'All this land in sight is very extensive and very high and does not seem to be less than continental. My God see fit that it is so!'[5] Quirós was even more certain: 'because the eye could not turn to a point that was not all land, the day was the most joyful and celebrated day of the whole voyage'.[6] What they saw were in fact overlapping islands.

At about five o'clock in the afternoon of 1 May 1606, the

three vessels were at the entrance of a bay described as 'very large and beautiful, and all the fleets in the world might enter it'.[7] The ships anchored for the night, and the next morning Torres left in a boat with twenty armed men to look for a suitable anchorage. People on the beach seemed to signal to the Spaniards to land, while three—in another account, two—canoes with some twenty archers approached the launch hesitantly, showing themselves 'to be troubled', no doubt alarmed and confused. Gestures meant to display friendliness had no effect, and as the warriors strung their bows, the Spaniards fired a few shots over their heads—or, in another record, those in the launch fired a culverin. The would-be attackers turned, rowing 'as hard as they could', and on reaching shore picked up their canoes and disappeared into the forest.

At about three in the afternoon the reconnaissance party returned to the ships, Torres and his men scarcely able to 'hold back the joyful news that they had found a good port', sheltered from the wind and with a clean sand bottom 8 to 30 fathoms deep close to shore.[8] 'Next day, being the 3rd of May, the three vessels anchored in the port with great joy, giving many thanks to God.'[9] The explorers would soon find that almost all the bay was extremely deep, and they had been fortunate to come upon one of its few possible anchorage sites. Quirós named the bay for Saint Philip and Saint James—San Felipe y Santiago—the 1st of May having been the day of the two apostles.

The bay so named by Quirós is now commonly called Big Bay. It curves deeply into the north end of the island of Espíritu Santo in today's Pacific island nation of Vanuatu. 'Like a horseshoe', Munilla wrote, some 15 leagues, he thought, in circumference. Wide black sand beaches were backed by a shadowy green wall of palms and shrubs and here and there great, spreading trees wreathed in vines, 'thick groves', said Munilla, 'in which by day and night numerous birds sang and it seemed as though we were in a delightful orchard'.[10] Westward a spine of forested mountains thrust its peaks against the sky. Several rivers emptied into the bay, one 'a stone's throw' from the Spaniards' anchorage, with a larger one about two leagues away. The humidity was high, but the temperature, cooled by the stir of trade winds, was 'so fresh that it obliges a man to cover himself with a blanket, a state of things to which we were not at all accustomed', according to Leza.[11]

To Quirós this was the 'so long sought-for, so good, and so necessary a port' granted him by God, which he would take in the name of the king.[12] He named the anchorage Vera Cruz and the rivers the El Salvador and the Jordan. Honouring Philip III's Austrian royal house, he named the land La Austrialia del Espíritu Santo, that is, Austrialia of the Holy Ghost, so the king might add this to his titles for the greater glory of God. Over the centuries this name has caused considerable confusion. Some subsequent transcriptions mistakenly used the word 'Australia', meaning South Land, and for this reason Quirós' name has sometimes been interpreted as a reference to what is now the continent of Australia. This, of course, is a mistake. Quirós repeatedly used the term 'Austrialia', which was derived from 'Austria', and had no reference whatsoever to today's Australian continent.

There were immediate necessities to consider. Having emptied their holds over the months of water, provisions and stores, the ships needed ballast. As well, closer investigation of the coastline with its river mouths, fringing reefs and shoals required the use of a smaller, handier craft, and the expedition was equipped to build a brigantine for this purpose. Inevitably fresh water and firewood were also needed.

Several days of heavy rain hindered these activities, but as the Spaniards reconnoitred from time to time in a boat, groups of natives threw them fruit and, laying down their bows, signalled them to land. Movements in the green bush behind them, however, suggested possible ambush and they did not land. By night there was the unceasing beat of drums and the sound of pipes and something 'like hawks' bells' emanating from the forest. That the land was well populated was obvious, inhabited right up into the mountains, according to Leza.

As the weather cleared, Quirós and Torres began searching the shores of the bay for a location suitable for the construction of the brigantine. Dense bush, however, grew near the water's edge. The labour of clearing an area for construction and for safety from attack would have been excessive, and the project was abandoned. Unfortunately, the lack of a light, manoeuvrable brigantine restricted and at times prevented the exploitation of their discoveries. For ballast, however, natural mounds of pebbles were found some distance along the beach. Wood and fresh water were close by.

Investigating the Jordan River, the Europeans were pelted with stones but refrained from retaliation. They needed peace to accomplish what was necessary, and over the next few days there were friendly exchanges. At times women and boys appeared, a sign that no violence was intended. Two men were colourfully dressed, soldiers assisting them with the unfamiliar garments. Nevertheless, the Spaniards were convinced that armed men waited in the thick vegetation.

On 9 May all efforts at friendship collapsed. Some 70 armed Spaniards, including arquebusiers and shield bearers, were landed under Torres' command to begin gathering stones for ballast and felling trees for a stockade and a shelter to be roofed with a sail. The boat was run up on the sand, the land proclaimed the possession of King Philip, and a cross erected.

The details of what happened next vary, but apparently a large number of islanders appeared from the bush, stringing their bows. Certain that more warriors remained hidden, the Spaniards formed up into a squadron. The two groups called to each other and made signs, which neither understood, the Spaniards, according to Leza, asking the others to lay down their arms. Three grey-headed men approached the Europeans. One of them, evidently of higher rank, drew a line in the sand, indicating that it was not to be crossed. They would lay down their arms if the strangers would do the same. Torres made it clear that he dismissed the idea—or in another account crossed the line. The warriors fell back 'skirmishing' with much clamour, arquebuses were fired and one native fighter fell. 'Our men hung the dead man on a tree', wrote Munilla.[13]

The fighting resumed, the Spaniards reportedly attacked by over 600 warriors. Quirós watched from his ship. Dismayed by the violence, he nevertheless felt it necessary to send in a support party of soldiers and fired two cannon so that the balls, tearing through the branches of the trees, passed over the engagement. At some point a chief was killed, and the attacks ceased. The gathering of pebbles continued and the stockade was erected. At the end of the day, however, Quirós ordered the stockade dismantled. The soldiers, in a warlike mood, obeyed with extreme reluctance.

In Prado's words, Quirós took the killings 'very ill'. It was the end of the peace he had sought in order to explore the 'grandeur' of this land he had discovered. Prado's view was very different: 'with such savages it was impossible to use politeness', and the action was taken 'in order that another time they should not be so rude to Spaniards, to whom all the nations in the world pay respect'. Of Quirós' reaction to these remarks, Prado added, 'He could not swallow it, being a Portuguese'.[14]

Prado's xenophobic arrogance was evidently not unusual at the time in a nation which had been for centuries under siege and suddenly cast by circumstance and enterprise in the role of conqueror. Such arrogance, probably especially common among the adventurous, militaristic types, conquistadors looking for conquest, was not, however, universal. Quirós saw all men, regardless of colour or culture, as God's creatures, with immortal souls that needed only to be given the blessings of Christianity. Torres was critical of some of Quirós' leadership, but worked smoothly enough with him. With the people of Taumaco Torres had established a rational relationship, but on Espíritu Santo he was at a loss. Of the people his description was flat and factual: 'They are all black people, and naked'. He continued: 'They did not choose to have peace with us, though we frequently spoke to them, and made presents: and they never with their goodwill let us set foot on shore'.[15] That the battle had hardened the inhabitants' will to resist and perhaps united separate clans or tribal groups against the invaders seems most likely.

The evidence suggests that at the time of the Spaniards' arrival, the local communities were in the midst of a celebration of their own, possibly engaged in an important chief-making ritual. Nights of drumbeating and piping, the fluctuating character of the reception given the Spaniards and the uncompromising rejection of strangers shown by the line in the sand seem to point to something significant being in progress. If this was the case, the intrusion of the Spaniards at this particular time would probably have been especially offensive. In addition, there was evidently no understanding or any effort at understanding on the part of the Europeans, who were instead extremely distrustful of even tentative amiability. There was good reason for distrust on both sides. The Spaniards' reception by island peoples could swing abruptly from an apparent welcome to a sudden attack. And whatever else the Spaniards might seem to be in the eyes of the islanders, they were inevitably dangerous, demanding and ultimately destructive.

SPE ET METV.

Chapter Fourteen
KNIGHTHOODS AND A CITY OF MARBLE

15 89

Following this altercation, the local people seem to have abandoned the area. Only by night were they seen on the beach, their fishermen's lights moving along the shore. By day the Spaniards entered deserted villages, which they found clean and well arranged, the houses with few exceptions liberally stocked with food, to which the soldiers helped themselves— coconuts, plantains, nuts tasting like those of Castile, some type of sweet citrus fruit, potatoes, yams and pigs. The soil seemed rich and fertile and in the cool streams, the men, well guarded, bathed and washed their clothes. The Franciscans came ashore, also impressed by the apparent richness of the land. Each night, however, by Quirós' order, everyone returned to the ships for safety's sake, and it was this that quite possibly prevented the spread of malaria, which had devastated the colony at Santa

Cruz. For some the question of the land's identity remained.
Munilla commented, 'We are in some doubt whether it is a
continent.'[1]

Quirós evidently had no doubts. The land enchanted him,
its mountain range and large rivers convincing him that it was
of great extent. Within days of his arrival, he was immersed in
plans for its administration and development. The safety of the
Spanish community was a priority, and he organised ministries
of war and marine which he made responsible for the protection
of shore parties, divine services held on the beach and the forest
trails they needed to use. He named nineteen men, the princi-
pals and their immediate subordinates, to these offices, among
them his nephew Lucas de Quirós as Royal Ensign, his friend
Pedro Lopez de Sojo as Captain and Sergeant-Major, Alonzo
Alvares de Castro, evidently a cousin, as one of two captains of
the infantry, Torres as Master of the Camp and Gaspar González
de Leza as Chief Pilot. In a lengthy speech he then explained
to all the chivalric order, the Order of the Holy Ghost, that he
was initiating and as a result, the new and greater responsibilities
that would rest upon each man for the 'discovery, pacification
and possession' of the land they were and would be exploring.
Ensuring a religious commitment to these goals, he had the
entire company make their confessions to Munilla and his three
priests.

The concept of chivalry as courteous and honorable behaviour would have been familiar in varying degrees to everyone
in the expedition. Chivalry, however, was also associated with
an aristocratic level of society, and applied to the rough-edged
majority of the men of the fleet must have seemed extraordinary,
if not incomprehensible. Quirós' thinking on this would have

been complex, however. He would have drawn from his awareness of the chivalric orders that emerged from the Crusades,
perhaps particularly the Templars with their headquarters on
the site of the Temple of Solomon, and the Hospitallers of St
John, also based in Jerusalem. Very likely there was in his mind
a link here to the city he now envisioned, a new Jerusalem. In
some sense Quirós might have felt that he had found a kind of
reality for the long-sought-after dream of mythical Ophir.

Saturday the 13th of May was the eve of Pentecost. About
nine o'clock that morning Quirós summoned Fray Munilla and
explained further the Order of the Holy Ghost, into which he
would initiate every person on the expedition, black, white or
Indian, including the islander taken at Taumaco. He had fashioned crosses of blue silk in varying sizes to denote rank and
favour, the insignia that all, the religious included, would wear
as Knights of the Holy Ghost. This proposal disturbed Munilla,
and he sent for the Franciscans on the *almiranta*. Meeting in the
stern cabin of the *capitana*, the friars concluded that knightly
insignia worn on their plain brown habits would suggest worldly
honours and vanity. They would instead wear wooden crosses.
In the late afternoon Munilla returned to Quirós with this
response. The commander was furious, erupting into one of his
abusive outbursts, a flash of the irascible Quirós of Lima. Munilla
left the cabin for his own quarters where, however, the agitated
old prelate could still hear the ranting of the commander, 'things
which cannot be set down with ink on paper'.[2] He went back
to Quirós and defended the Franciscan habit and all it stood
for—it required no knightly insignia.

Meantime a shore party was building a chapel of tree branches
and plantain leaves. The launch was anchored nearby so that its

artillery could be used in conjunction with the festivities. In the evening there was music and dancing, bells were rung and drums beaten and, as Leza recounted, 'a great festival of rockets, fire-wheels, and firing off the guns, which caused a great echo on the land', the sound seeming to roll down from the mountains.[3] From the forest the Spaniards could also hear the shouts and cries of the natives, their ears assailed by the thunderous, inexplicable noise.

Before dawn the following morning Torres and the new ministers went onto the beach with an armed party, landed four small cannon and erected stakes to protect the leafy little chapel, named for Our Lady of Loreto, from possible attack. Quirós and the rest of the expedition then came ashore to walk in procession between the drawn-up lines of the three ships' companies, each man wearing his blue silk cross. There were, as Munilla remarked, sailor-knights, grummet-knights, ship's page-knights, Indian-knights, knight-knights and others. As Royal Ensign, Lucas de Quirós carried Spain's red and gold standard. Quirós went down on his knees. 'To God alone be the honour and the glory,' he said, and kissed the ground. 'O Land! sought for so long, intended to be found by many, and so desired by me.'[4]

A cross made from local wood was embraced by the kneeling, bare-footed Franciscans. Rising and singing, they carried it, leading the procession to the entrance of the chapel, where they set the cross in place. Guns were fired in a rolling salute of twelve salvoes as the Blessed Sacrament was carried past lowered flags and standards into the chapel. Quirós spoke in a loud voice on the significance of this cross and read out six long statements whereby he took possession in the names of King Philip III, Jesus Christ, the Pope and several saints 'of all the islands and lands I have newly discovered, and desire to discover, as far as the South Pole'. This pronouncement, re-enforcing the claims of Balboa almost a century earlier, confirmed what Spain saw as its entitlement to the entire Pacific region, a claim the monarchy would never formally revoke.

Quirós then went on to announce that Spanish administration practices would be introduced,[5] and here in the latitude of 15° 10' South he would found a city to be named New Jerusalem.[6] There were resounding and emotional cheers for the king. Four masses were said, the flags and banners blessed and the sacrament fervently taken by all.

On the ships, where banners and pennants at the mastheads streamed in the wind, all the guns were fired, rockets and fire-wheels set off, and muskets and arquebuses discharged. Ashore, two men who each owned a slave agreed to free them. 'This being done', Quirós wrote, 'we went to dine under the shade of great tufted trees near a clear running stream, the *corps de garde* being alerted and the sentries posted'.[7]

Not everyone was impressed. Prado, apparently a knight of the old and prestigious Order of Calatrava, said scornfully, 'It was all wind, both walls and foundation', and took the opportunity to scathingly demand of Quirós the gold, silver and hatfuls of pearls he had promised. According to Prado, the angry commander did not reply. In his journal the accountant Iturbe also ridiculed the order and its blue taffeta crosses, which caused considerable amusement, he said, when even 'two negro cooks were knighted and rewarded by such largesse, great liberality and munificence for their gallantry and courage'.[8]

After his customary afternoon rest, Quirós summoned his chief ministers and formed a municipal council for New

Jerusalem, now the capital city of a Spanish province. The lists of
names and positions recorded do not agree in every detail, but
essentially there were thirteen magistrates, a secretary, justices
of the peace and royal officers who included an accountant, a
treasurer, a registrar of mines and others. Quiros' plans for the
city of New Jerusalem were recorded with heavy sarcasm by
Prado:

> its gates were to be of marble and he pointed out that it was to
> be got from a white spot that was in a clay pit about two leagues
> off. And the great church was to be of that marble and was to
> be such as to rival that of Saint Peter at Rome, and fences of
> the city and houses also were to be of that marble . . . and he
> was going to write to his Majesty to send him three thousand
> friars to plant the holy catholic faith therein; and other things
> very tedious to relate.[9]

One cannot know to what extent Quirós might have drawn
upon the prophetic biblical vision of a wondrous New Jerusalem descending from heaven. Interestingly, it was the cynical
Diego de Prado who some years after the voyage became a
monk.

The day's affairs concluded, Quirós embarked for his ship
where, as he arrived, he ordered the execution block taken
down. He could no longer believe that 'persons with such an
honourable destiny would do things the punishment of which
would be the rope'.[10] Those remaining ashore faced another
problem. Apparently the feasting had all but exhausted the
expedition's provisions. Thus, with 80 arquebusiers and shield
bearers and accompanied by Fray Munilla and three other

religious, Torres now went in search of food. Following the sound of revelry farther inland, they were briefly confronted by armed warriors, but then came upon the site of a hastily abandoned feast—quantities of yams, four tied pigs, as well as meat and other foods roasting over a fire. Munilla wrote that the new knights promptly made off with the spoils.

The new land, however, needed to be explored. Repairs were made to the ships, and Leza came ashore to observe the variations of his compass, take the sun on solid ground and adjust his astrolabe. Well-guarded fishing parties left with their nets before dawn and hauled in large quantities of fish. Sweet fresh water was obtained from the nearby stream of El Salvador, while the launch and the boats investigated the Jordan River. The men saw farms and a village from which the inhabitants fled as the vessels approached. Prado produced a map of the bay.

How the surveying was done is not described, but probably it was carried out by Torres or his pilot Fontidueña, and possibly Ochoa, with distances estimated by eye and compass bearings taken from the ships and from the boats as they were rowed along the shore. Under these circumstances Prado's result was exceptional, a combination of plan and perspective, with minutely drawn trees, rivers, headlands and anchorage sites neatly captioned with Spanish names. There are four surviving maps made by Prado, each of a harbour where the expedition remained for some time, the first at Espíritu Santo, the others on the New Guinea coast.

Torres led several excursions inland, sometimes penetrating several leagues into the country, seeking food as well as some knowledge of the land. They came upon numerous huts and cultivated plots, and a village of more than 100 dwellings. The inhabitants were rarely seen, diappearing into the forest on the strangers' approach. In one instance a woman with a small boy in her arms was taken. She thrust the child at Torres and vanished into the thick vegetation. The Spaniards baptised the boy and left him for her to find in one of the huts. In an altercation not explained, they killed two men. On another occasion Torres and some 40 men climbed a high, steep mountain and found themselves looking down upon a beautiful plain where, descending, they discovered fruit trees, nutmeg and nuts similar to almonds. 'Here one might live in great luxury,' commented Leza.[11] They were attacked but retaliated, capturing, as Leza noted, 'three beautiful women; but we let them go, because our General [Quirós] did not wish any woman to be brought on board. We also caught three boys'.[12] On the trek back to the shore, heavily laden with provisions, the group again came under attack, 'the barbarians shooting many arrows and darts at us from amongst the trees, but we got no hurt, nor did we have to drop any of the loads'.[13] On the ship not everyone was pleased with the task of overseeing three children. Someone spoke up in a loud voice: 'Thirty pigs would be better eating than three boys.' This elicited from Quirós a speech on the infinite value of saving those three young souls.[14]

Quirós himself made a brief foray inland. With an escort of soldiers he went to a deserted farm and liberally sowed maize, cotton, onions, pumpkins, melons, pulse, beans and 'other seeds of our country'.

On 25 May the Feast of Corpus Christi was celebrated. Very much as Pentecost had been observed, there were gun salutes, masses in a chapel refurbished with new green boughs and, in a solemn procession under palm-frond arches, the officials of the city of New Jerusalem carried the poles of a canopy held over the Holy Sacrament. After the afternoon rest there were dances in the 'Portuguese fashion'—a sword dance, with eleven sailors in red and green silk and with little bells on their feet performing to the sound of a guitar; and a dance by eight garlanded boys, singing in praise of the Holy Sacrament to the accompaniment of flutes and tambourines. Quirós recorded that 'having given the souls such sweet and delicious food', the company headed to the cooking sites and 'gave themselves up to feeding their bodies'.[15]

On many the emotional impact of the day was deep. For a small company of men infinitely far from home, isolated and surrounded by hostility, the support of their beliefs, the reassurance that what they were doing was good and valuable, and the camaraderie of a good meal together would have been profoundly cheering. Quirós' secretary, young Bermúdez, wrote movingly of the ceremonies and adapted a verse from the epic *La Araucana* to the occasion. The young man taken from Taumaco, now named Pedro, was also dressed in silk, wearing his blue cross and carrying his bow and arrows. Someone with unusual understanding had allowed him to preserve his dignity as a warrior. Quirós concluded the day with a march inland with 100 arquebusiers, 30 shield men and the drummers, and had the satisfaction of seeing that the seeds he had sown earlier were sprouting.

There was much more territory to explore, and the next day preparations for departure were under way. There was some successful fishing, fresh water was taken on board, the men bathed and washed their clothes, and a final foraging expedition was made. There was, however, several days' delay when fish poisoning struck, particularly aboard the *capitana*. The *al-miranta*'s surgeon administered oil, which produced vomiting and recovery.

Three days later there appeared on the beach a huge crowd of natives, creating more commotion than the Spaniards had seen before. It became clear that they were demanding the return of the three boys taken several days earlier. Sign language negotiations for the exchange of the boys for pigs commenced, together with an attempt to trade two goats for pigs, all of which failed. In the end the Spaniards acquired one pig but surrendered none of the children. It seems possible that Quirós had no intention of releasing the boys; his initial price for each boy was an impossible 30 pigs. They would become Christians and in time return to their homeland to assist in spreading the faith. To him, the importance of this transcended the obvious grief of the children and their parents. Munilla wrote that the islanders were considered 'despicable' for not wishing to give more than one pig for each boy, but carefully tended for 10 to 20 years, the tusked pig was an extremely valuable and possibly a sacred animal.

On the morning of 8 June, with the ships' companies restored to health, the fleet made sail. Positive traces of the Spanish encampment at Big Bay have not been found nor the exact location established, despite such claims being made from time to time.[16]

SPE ET METV.

Chapter Fifteen

SEPARATION

15 89

Quirós intended to sail towards what appeared to be a high blue mountain range lying across the southern horizon, initially keeping the bay and its anchorage to leeward should a return become necessary. An easterly breeze carried the vessels to the entrance of the bay, but as they moved into open sea the wind veered to the southeast, rising in intensity. Quirós ordered that they continue, and under courses only the vessels rolled and plunged in gales and mounting seas. They made no headway, and on the evening of the 9th of June Quirós ordered a return to the bay. This could not be readily achieved. The vessels laboured, tacking endlessly, until on the late afternoon of the 11th the *almiranta* and the launch made it into the bay and by nightfall were secured near their previous anchorage.

Lanterns were lit to provide leading lights for the *San Pedro y San Pablo*, which was finally moving into the inlet, and by one report came close enough to the *San Pedro* to be made out in the heavy darkness, and by another to hear the men taking in sails and, from the chanting, lowering the anchor. According to Prado, aboard the *almiranta* they signalled the flagship with a torch at nine o'clock and received a reply. They signalled again at twelve, but there was no response.

At this point the several narratives concur in general but vary in detail. At seven o'clock, or by another account at nine that night, a fierce wind swept across the bay, and the *capitana* was suddenly in danger. With topsails still set, the vessel listed perilously until, according to Munilla, ropes were untangled and the sails lowered. On taking soundings, the *capitana* could find no bottom. Lights were seen, but her people could not determine whether they were those of the other ships or the torches of native fishermen.

As the ferocity of the wind mounted, according to Leza, Quirós and his officers decided to stand for the middle of the bay. Pedro Bernal Cermeño had recently been promoted to admiral, a ranking officer usually in command of his own ship, but in this case remaining aboard the flagship. According to Leza, now chief pilot, it was then 'resolved by the same persons' to strike the topmasts and run before the wind under a spritsail to find shelter behind a promontory outside the bay. Leza makes no mention of himself. Yet during the investigation conducted in Mexico the following year, Cermeño declared that the responsibility of navigation that night was Leza's, not his. What role then did Cermeño play as admiral? To what extent did Quirós share in the decisions? Bermúdez, writing on Quirós' behalf, says he was 'disabled by illness'. Thus the question: who was actually in command? From statements made at the enquiry, there seems to have been considerable confusion on this point. Iturbe, who had no part in navigational matters, was nevertheless sharply critical of the decision to leave the bay:

> [The ship] could have spent the night tacking to and fro, because it [the bay] is broad and free from rocks, or they could have anchored . . . But nothing would satisfy them except to return to the danger from which they had fled, leaving their companions at anchor.[1]

To whom this criticism was directed is not clear. Bermúdez, again in Quirós' journal, went on to say that 'more diligence' could have been used in getting the ship to anchor within the bay, and remarked rather astonishingly that 'it was also said that they went to sleep'.[2] Bermúdez did not explain who 'they' were.

Whatever the situation, at dawn the ship was found to have drifted to leeward of the bay entrance, by most of the accounts 20 leagues or more. The weather worsened. For three days the *San Pedro y San Pablo* struggled to return to the inlet. On 13 June Leza wrote: 'I took the sun in 12° South, so far had we been taken to the N.W. . . . [unable] to turn her stern from the wind lest she should broach to'.[3]

There was no sign of the other vessels. Quirós now resorted to his original directive, issued at Callao, that if separated the ships should rendezvous at the island of Santa Cruz. Apparently Quirós met with the friars and several of the 'responsible and honourable' people on board to whom he explained the option of making for Santa Cruz. According to Iturbe, Quirós then left the group 'and went up to where the pilots were and issued orders to change the course and to head for New Spain', that is, Mexico.[4] Munilla, however, records that there were several days of sailing in search of Santa Cruz before the change was made. Thus exactly when the decision was made to steer for Mexico remains uncertain.

Prado provides a more dramatic version of what brought about the change, which he said he learned later from sailors from the *capitana*. According to Prado, the threat of mutiny had finally become a reality. When the *San Pedro y San Pablo* failed to re-enter the Bay of San Felipe y Santiago, Quirós was deceived into thinking they would shelter temporarily behind a certain cape and retired to his cabin to sleep. Emerging at midday, he found the ship headed for Mexico. Supposedly, a plan to throw him and his nephews overboard was dropped and the three men, together with their friend Sojo, were kept under guard throughout the voyage. Such was the story recounted years later in Prado's *relación*. The testimonies of those actually on board the *capitana* at the time, which include sworn documentation, record no such events. From the evidence other than Prado's there was no mutiny.

What, then, caused Quirós to change course and destination? By so doing he would cut short his explorations. The flagship would miss meeting the *almiranta* and the launch if, following instructions, they eventually reached Santa Cruz. There were, however, difficulties. When earlier in the voyage the *capitana* continued westward on the correct latitude for Santa Cruz, the explorers realised that they did not know whether the island lay to the east or the west of their position. The possibility of making for Manila was raised, but at midyear the monsoonal winds would have been against them. As was customary, the expedition had originally received provisions for one year, on the assumption that after six months a ship would turn for home. For Quirós' expedition that six months had just expired, and it appears obvious that more than half the provisions had been consumed. As well, there was water for only three months. Perhaps to Quirós the most compelling reason for returning to the Americas was that he believed he had found Terra Australis. His reconnaissance mission was complete. Bringing the great news back to the king was of first importance and certainly an attractive incentive.

Not everyone agreed. On or before 16 June, Iturbe had learned of Quirós' intention to abandon the voyage and head for New Spain. He promptly drew up a letter of 'remonstrance', which he gave to the notary, Juan de Arano, to deliver to Quirós. He cited the *cédulas* by which the king had commanded that discovery take place. An effort had to be exerted to meet the other vessels at Santa Cruz, and an attempt made to sail to New Guinea or the Solomon Islands in order to present the king with the report of a more complete voyage.

Perhaps concerned by the obvious discontent of some of his men, Quirós on 18 June 1606 called a meeting of his officers, officials, gentlemen and seamen, at which they were to state their opinions on what should be done, to be formally recorded by the notary. Cermeño, as admiral, spoke first. He believed the ship could reach New Spain in three months; the vessel was in good condition and had sufficient stores. Successively the others deferred to his experience, although Bermúdez expressed a wish to continue exploration. Evidently the mystique of the Great South Land still held the imagination of the young poet.

Strangely, thc namc of the veteran *entretenido,* Alonso de Sotomayor, was omitted from the list of those making their statements. Sotomayor's resentment against Quirós' actions did not surface until 1613, when he wrote to the king, listing Quirós' failures and suggesting 'severe punishment' for him. In 1606, however, Sotomayor was evidently silent.

Quirós read the men's statements and issued his orders. The course northeast by north was to be continued as far as Guam and the situation then reassessed. This was confirmed on 21 June, and Leza recorded the event. On the 22nd, however, he wrote two sentences: 'We went 25 leagues to the north. On this day we saw many birds'.[5] And for the next three months, apparently registering his disapproval, the pilot's daily entries typically consisted of a single sentence or less.

The *San Pedro y San Pablo* crossed the Equator on 2 July 1606. On 23 July at 13º North, close to the latitude of Guam, Quirós conferred with Cermeño, Leza and Francisco Fernández, the second pilot, on changing course for Manila. Cermeño reiterated that contrary winds were to be expected on sailing to Manila at that time of year, whereas minimum sailing time could be expected on a course for Acapulco. Leza and Fernández agreed, and Quirós ordered the pilots to steer for the coast of New Spain and the port of Acapulco. Too ill to be in the deck house, he kept Cermeño near him to pass on his orders and give him a daily report on matters on board. On a heading roughly northeast and then east along the 38º North parallel, the solitary galleon made its way across the emptiness of the North Pacific. At 34º North the coast of New Spain was sighted, approximately where the modern city of Santa Barbara, California, now stands, and steering southeast the ship followed the shoreline.

The privations of the long journey had been many. Half their fresh water was lost because the jars were not properly sealed with pitch, and what remained became fetid. Bread deteriorated. Only rain and successful fishing kept the men alive. Within sight of the California coast a man disappeared, apparently attempting to reach land on a crude raft. In late August the aged Fray Munilla made his last entry in his journal, and the record was taken over by Fray Mateo Vascones. There were quarrels which, due to what Vascones called 'poor command', were not settled. Quirós, in fact, remained in his cabin, reportedly for three months, confined by what many believed was a pretence of illness.

On 11 October, crossing from off Cape San Lucas to the Mexican mainland, the ship was gripped by a terrifying storm, massive seas sweeping over the decks, men waist deep in water, others huddled in corners waiting to die. Thunder-claps seemed to split the sky and there was 'such a smell of sulphur that it seemed an infernal thing'.[6] Dwarfed by towering seas, 'there seemed to be nothing left for the ship but to turn over on her keel', said Vascones.[7] But she survived. On 13 October, with the passing of the storm, Fray Martín de Munilla died, at nearly 80 years of age. On 21 October the *San Pedro y San Pablo* reached the sparsely inhabited inlet of Navidad in Mexico and on 23 November 1606 entered the port of Acapulco.

Chapter Sixteen
THE LAST JOURNEYS OF QUIRÓS

S hortly after his arrival in Acapulco, Quirós was ordered by the Viceroy of New Spain, Don Juan de Mendoza y Luna, Marquis of Montesclaros, to surrender the *San Pedro y San Pablo* to royal treasury officials. On receipt of a second order, Quirós complied, an inventory was made, and on New Year's Day 1607 he set off for Mexico City. Five days earlier the viceroy had written to the king, citing the complaints against Quirós received from the ship's pilots, soldiers and friars. In Mexico an enquiry was held and witnesses submitted testimonies. However, no action was taken against Quirós and, assisted by the generosity of friends, he left New Spain and on 9 October 1607 arrived in Madrid.

Quirós would spend the next seven years seeking royal support for a colonising expedition to what he was certain was the mainland of Terra Australis. Imbued with missionary spirit, he now believed with absolute conviction in the spiritual nature of his mission and that to accomplish it was his destiny. Virtually on his arrival in Spain he began writing a long series of memorials, petitions addressed to the king, which he had copied or printed and distributed among members of the Councils of State, War and the Indies, and the royal ministers, claiming later that he wrote 50 in 50 months. They were panegyrical descriptions of the lands he would discover, lists of requirements for the voyage, his plans for colonisation. Quirós saw his coming to Terra Australis as a spiritual conquest, but to win the interest of his king he also described the riches the land would produce.

Eight memorials are preserved, the first, written in 1606, describing the events of the journey and avoiding blame for altering the fleet's course at 26° South and for parting with Torres. It also contains the explanation for the name Austrialia del Espíritu Santo. He wrote of his financial hardships: he had had no pay and was in debt for 2500 dollars. In a later memorial Quirós compared himself with Columbus, Da Gama and Magellan. By late 1610 the Council of the Indies had become concerned with the widening circulation of Quirós' memorials, and the possibility that important information might be falling into undesirable hands. The king finally ordered that Quirós gather his memorials and hand them over to a Council official.

The Council's possible interest in Quirós' project was diminished by his very considerable requirements in men, ships, equipment and money, and his demand that his authority be unconditional and documented as such. Quirós probably did not understand the exhausted state of the royal treasury and the problems of an already overextended and largely indefensible empire. Similarly, he might not have known of the numerous letters still reaching the court criticising his leadership on the 1605–07 voyage, among them two particularly vicious missives from Prado. The 1613 letter of Alonso de Sotomayor was similarly censorious.

At the same time the Council of State was becoming concerned that this fanatical mariner, if desperate enough, would sell his knowledge to Spain's enemies. He needed to be appeased with promises, and may have been kept occupied for a time with an appointment as a cosmographer. The Council also questioned whether it could 'with good conscience ... make these conquests of heathens who neither disturb nor attack us'.[1] However, in 1614 it was finally agreed that Quirós would return to Peru with the newly appointed viceroy, Francisco de Borjay Aragón, Prince of Esquilache, who would implement the voyage of discovery when opportune. With this, the business of 'this man' was finally considered finished.[2] There appears to have been no formal directive to the viceroy. On the contrary, a private document apparently urged him to keep Quirós diverted, but not to give him an expedition. Exhausted by years of effort and unaware of the secret instructions, Quirós accepted what he was offered.

In early April 1614 Quirós, with his wife, son Francisco, daughter Jerónima de Alvarada, five servants and possibly his secretary Bermúdez sailed from San Lúcar de Barrameda with the viceroy's fleet, arriving in Panama in June. No passage money was charged. The viceroy's entourage reached Manta in Ecuador on 9 September and Lima on 10 December, but Quirós was not with them. A document of 1615 states that he died in Panama,[3] but no details are given. The viceroy's correspondence with the king makes no mention of Quirós, but 400 pesos remaining of the funds allotted to Quirós for the journey were passed on to his widow. Quirós' son Francisco apparently continued on to Peru and in subsequent years held several government positions. His nephew Lucas acquired the title of 'Cosmográfo del Peru'. Two of his maps survive. No further word of his daughter Jerónima has been found. Bermúdez returned to Spain in 1616, where he pursued a successful literary career. Several of his plays and poems are extant.

Chapter Seventeen
A DIFFERENT COMMANDER

O n the morning of 12 June 1606 the fierce winds that had swept out of the South Pacific Ocean during the previous days abated and relative calm settled over La Austrialia del Espíritu Santo. The men of the *almiranta, San Pedro*, scanned the Bay of San Felipe y Santiago for the fleet's flagship. She was nowhere to be seen. Torres took the ship's boat along one shore and sent the launch along the other, the men watching for wreckage or any other sign of the *San Pedro y San Pablo*. Landing, they climbed the headlands to look seaward, but saw an empty ocean. Prado assured everyone that a mutiny had taken place.

Apparently not convinced, Torres waited. Among the many disadvantages of the *capitana's* loss was the fact that the ship had carried the expedition's trade goods, tools, a good part of its medical supplies and 'many other things', as Torres noted. After fifteen days Torres called a council of his officers and those of the launch, probably gathering in his cabin at the stern of the ship. Here he brought out the sealed orders received from the viceroy of Peru on sailing from Callao, which named a successor to Quirós in the event of his death or incapacity. No copy of these orders survives, and no official document that refers directly to the contents appears to be extant. Hence there is only speculation as to who was named commander. In his narrative of the voyage written several years later, Prado stated that he was designated to succeed Quirós, but there is substantial evidence against his claim. That Prado received the position is recorded only by Prado himself—and this several years after the purported event. Further, he subsequently refers several times to Torres as *cavo* or commander, and two contemporary documents do likewise. At the end of the journey it was Torres, not Prado, who sent the report and maps to the king, the duty of a commander. In his later letter to Philip III, a relatively short, straightforward and factual account of the voyage, Torres makes no mention of Prado. Prado had previously demonstrated his self-importance. Was this the role in which he wanted to be seen by posterity? Or had Prado been appointed captain of the flagship, which, of course, was gone, an appointment he chose to interpret otherwise? Without a copy of the viceroy's orders there is no answer other than what can be drawn from the events that followed.

The action to be taken now was discussed. Philip III's instructions were to sail to 20° South latitude in search of Terra Australis. Clearly there was opposition to the idea. Later, Torres wrote to the king:

> I brought forth Your Majesty's orders . . . it was agreed that
> we should fulfil them, though against the inclination of many,
> I might say the majority, but my condition [disposition] was
> different from that of Captain Pero Fernández de Quiros.[1]

This statement leaves little doubt that Torres had assumed command. The new commander had also established his authority and was obeyed.

At this point Torres would have assessed his situation. The *San Pedro* was in good condition, but the provisions brought from Peru were rotting. On board he had two pilots, Fontidueña and Ochoa, the latter deposed from the flagship but evidently creating no difficulties for Torres, and as master Gaspar de Gaya (or Gaes). There were three Franciscans, including the chaplain Fray Juan de Merlo, as well as Francisco, the Peruvian lay brother, and two brothers of St John of God. He also had two surgeons, one of whom had tranferred from the *capitana*, and two *entretenidos*, including Prado, whose narrative of the voyage would become its principal record. Evidently there were adequate numbers of sailors and soldiers with the necessary mates, gunners and sergeants.

The launch *Los Tres Reyes* was commanded by Gaspar Gonzalez Gómez. Its company included a master-gunner, a captain of the infantry, and his sergeant and twelve seamen. That Gómez handled the little vessel competently is obvious. He met with no hardships that were of his making and remained in company with the galleon throughout the journey.

What charts did Torres carry? We have no record of this. It can be assumed that the earlier discoveries of Mendaña and Quirós were inked in on whatever maps he had, and that the north coast of New Guinea, the Moluccas and at least some of the Philippine Islands appeared. A later remark by Prado suggests that at some time he had seen a map that showed New Guinea as part of a mainland extending to the antarctic pole, but neither its source nor any detail is mentioned.

In preparation for the continuation of the voyage an attempt was made to obtain pigs for fresh meat, but the opposition of a large number of native warriors prevented this. The seamen obtained wood and water, and Torres wrote, 'I set out from the Bay in fulfilment of the order'. First, however, he sailed far enough around La Austrialia del Espíritu Santo to confirm that it was an island, not part of a mainland. Then he evidently steered the *San Pedro* and the launch toward the southwest. He continued:

> I had at this time nothing but bread and water, and in the height of winter, and the sea and contrary wind and ill wills: all this was not strong enough to prevent me from reaching the latitude of which I passed one degree, and would have gone further if the weather had permitted, for the ship was good . . .

At 21° South, he was well into the Coral Sea, probably some 300 kilometres from the Australian continent. Some drifting logs and reef birds were seen but 'I did not find therein a sign of land'.[2] Firm leadership had carried the men despite their objections.

Failing to find land at 20° South, the vessels were to proceed to Manila and to wait there for four months for any part of the fleet that had become separated. Then, refitted and reprovisioned, they would sail for Spain by way of the Cape of Good Hope. Accordingly, the galleon and the launch headed north, intending to round New Guinea's east end and, after following the known north coast to the Moluccas, swing northward again for the Philippines.

In the semi-darkness of early dawn on 14 July a cry from the lookout brought the men rushing on deck to see before them a terrifying white line of breakers. At 11½° South they had fallen 'in with the beginning of New Guinea', the Louisiade Archipelago, a tail of reefs and small islands stretching east from the New Guinea mainland and unknown to Europeans. Probably this landfall was Tagula Island. Unable to clear the reefs and islands against fresh easterly trade winds and the ocean current that they created, Torres put about and steered west along the uncharted island chain.

Torres' immediate objective was the Moluccas, the Spice Islands, now Indonesia's Maluku. Two questions can be asked: did Torres believe that by steering west along the New Guinea south coast he would emerge in the Moluccas, and had he considered the hazardous possibility that he was entering a bight where he could be embayed indefinitely by easterly winds?

New Guinea had been depicted as an island on the world maps of a number of cartographers, among them Rumold Mercator (1587) and Cornelis de Jode (1590). These maps, however, included legends to the effect that it was uncertain whether New Guinea was insular or part of Terra Australis. On such maps the strait between New Guinea and the southern continent is laid down at about 18° to 22° South latitude, when in fact it lies roughly between 9° and 11° South. Thus its placement was purely hypothetical. Torres would have been aware of the debate on New Guinea, and doubtless Prado shared with him his recollection of a map showing New Guinea at one with land to the south. What thoughts Torres may have had on this problem we do not know. The fact was, however, that with strong, persistent winds from the east he had little choice but to steer west.

For five days he cautiously followed the Louisiade chain, steering out to open water each evening to remain under shortened sail through the night. The archipelago gave way to a long, low shoreline to the north, here and there edged with pale green shallows or murkier water around the thickly interlaced stems of mangroves.

On 18 July, with the ship's boat ahead sounding, the galleon and the launch entered a sheltered inlet with a clean bottom at 14 fathoms, and ashore, springs of fresh water among large trees. Here Prado drew his second map, naming the bay Puerto de San Francisco (later Puerto de Lerma), which appears to be today's Sukuri or Oba Bay. A larger bight also depicted is now Milne Bay. Prado's maps are important in tracing Torres' route. Dates of arrival and observed latitudes were recorded, and are less subject to error than those in narratives written or rewritten months or years later.

The next morning, an excursion ashore sent the inhabitants fleeing into the hills while the Spaniards raided yams and potatoes from rush-fenced gardens and glimpsed the sea on the other side of where they stood. The following day the two ships' boats began searching for a way through, but were met at the mouth of a narrow sea passage by six large canoes with twenty paddlers to a side, in full battle array with flying streamers, gesticulating and shouting. The Spaniards opened fire.

The warriors fled, but regrouped on shore with round shields, clubs and wooden swords. Again the Spaniards fired. Of their opponents Prado wrote, 'when any fell dead they gave them blows with their clubs to make them get up, thinking they were not dead'.[3] Briefly the fighters were joined by some 300 others, but having spoken with the original group they then assisted with heaving the dead onto their shoulders, the entire force silently running off.

The Spaniards made their way through the passage, which they named Boca de la Batalla (Battle Passage, now Sawa Sagawa), to find two adjoining harbours, the bigger one today's Jenkin's Bay or Liliki, beyond which they saw the large expanse of Milne Bay. The men spent three days exploring the area. The return through the sea passage was harrowing. It was narrow and rocky, the ebbing tide running through so powerfully that on meeting wind-driven waves from the other side, huge breakers sprang up. The boats were hauled through by ropes. They had, in fact, found a route that would lead to open sea, but it was narrow and filled with rocks and rough water. Torres evidently decided to take no further risks.

On 1 August a large shore party took formal possession for Philip III with cheers, salvos and in the evening 'illuminations' and the singing of the Salve Regina. Interestingly, Prado claims that he conducted the ceremony of possession, but on his map credits the discovery of the area to Torres. After fifteen days in Puerto de San Francisco the two vessels resumed their journey west. They were well provided with food, which included pigs and some very large fish, parts of which were packed in salt brought from Peru.

In the next eight days they followed the coast for about

95 kilometres, sailing only by day, with a lookout at the bowsprit. Torres mentions large rivers and inlets, which presumably they investigated. On 12 August they dropped anchor in the shelter of an island, probably Bona Bona, which they named Santa Clara. Some natives were captured and ransomed for a 'fine big pig'.

From Bona Bona the ships' boats explored the mainland to the east. On 15 August there was an extraordinary meeting at a large upriver village where, apparently unafraid, the people welcomed the strangers and the two groups sat on the ground together. Asked what they wanted, the Spaniards indicated 'water', which was brought to them in a large cane tube. When a soldier asked for a big pig that had wandered in among the men, it was given to him. He put the match to his arquebus, aimed and shot it in the head. This caused 'great astonishment', and the man who had brought the water asked for the gun. Going up to a very fat pig, he aimed and loudly called out, 'Pu!' When nothing happened he aimed again, shouting more loudly. His companions were now roaring with laughter. The soldier, taking back his arquebus, turned aside, recharged it, and killed the pig, whereupon the laughter was 'still greater'. There was an exchange of a 'Milanese' bell on a silk ribbon for a large bird.

As they did elsewhere, Prado and more briefly Torres described the people they met, their dwellings, domestic animals, wild life and vegetation. Their descriptions of skin colour are sometimes confusing for, in making comparisons with the skin tones they knew from the racial mixtures of South America, they described some natives as white or not white, black, mulatto or in similar terms.

Prado's next map provides the evidence for Torres' subsequent route. Crossing the broad expanse of Orangerie Bay, so named in 1768 by Louis-Antoine de Bougainville, and skirting shoals and islets, the vessels appear to have anchored close to an island Torres called San Bartolomeo, Mailu on modern charts. Here the landing party found that the path to the village ran through a narrow, heavily guarded defile between precipitous mountain slopes. Signs of peace were answered by the defenders with shouting and the brandishing of lances and shields. Further peace signs had no more effect and Prado wrote:

> Seeing that we were losing time by treating them with further consideration we knelt down and saying a Pater Noster and an Ave Maria, Cierra España [the ancient Spanish war cry], we gave them a Santiago [an attack invoking Saint James] . . .[4]

The defenders fled, escaping in 26 canoes, with the Spaniards firing after them. That they could be killed at sea seemed to horrify the warriors even more. Women, children and the elderly, however, had retreated to the top of one of the heights above the defile. The Spaniards signalled the people to come down, and were answered with a shower of rocks. Ochoa and a Galician volunteered to climb the cliff face, but halfway up were so heavily stoned that they 'came tumbling headlong to the bottom'. In response to this, over twenty soldiers with shields forced their way up and 'made slaughter'. 'I was sorry to see so many dead children', commented Prado, who selected fourteen boys and girls between six and ten years of age and sent them to the *San Pedro*. One girl of about fourteen was so attractive that there were disputes among the men as to who was to take her to the galleon. Fearing that 'some might fall away with her and offend God', Prado restored her to her people. The children taken aboard were put under the tuition of the friars. Later, on reaching Manila, they were baptised, 'to the honour and glory of God'.[5]

The expedition left Mailu on 28 August, steering west along the outer edge of a barrier reef that bordered the coast. Landward there were forested hills and cloud-wreathed mountains. Two days later they apparently found an opening in the reef and warily approached the prominent 183-metre headland of Taurama and the island of Manubada, where they anchored off what is today Papua New Guinea's capital city of Port Moresby, at that time a busy centre of native trade. Again at sea, a fierce squall descended, the dark, wild world around them slashed with lightning and resounding with thunder. A cable broke and an anchor was lost, and the vessels ran under bare masts until the winds abated. The heat of the day was lifted, but humidity hung heavily.

Torres now faced the expanse of the Gulf of Papua, curving deeply into the land and some 400 kilometres across where it edged the Coral Sea. It had been seven weeks since his first landfall in the Louisiades and he had covered almost 1000 kilometres, exclusive of detours in and out of anchorages and out to sea for the night.

Different writers have espoused two principal routes possibly taken by the *San Pedro* and *Los Tres Reyes* in traversing the Gulf, for both of which a few words of support can be gleaned from the narratives. Neither account is clear, however. The vessels could have followed the arc of the gulf past the deltas of New Guinea's great rivers. If so they would have skirted the huge sandbanks of the Fly, flowing 1200 kilometres out of the highlands. Similarly, they would have carefully avoided the jigsaw-like delta of the Kikori and the swamplands through which the Purari wound its serpentine course. Or on a direct westward heading, the two vessels could have cut straight across the gulf from a point north of Port Moresby to Parama (Bampton) Island off the western shore. The argument is complicated by small but possibly significant variations in the existing translations of Spanish records. Whichever was Torres' choice, his objective was to reach Manila, which he knew lay in a northwesterly direction in approximately 14° North latitude.

Having worked his way westward this far along the New Guinea coast, he may have thought that his route would now be fairly direct. But on 5 September he wrote that their course had brought them 'into 9°' and:

> we fell in with a bank [*placel*] of from 3 to 9 fathoms, which extends along the coast above 180 leagues . . . We could not go farther on for the many shoals and great currents, so we were obliged to sail out S. W. in that depth to 11° S. latitude. There is all over it an archipelago of islands without number, by which we passed, and at the end of the 11th degree the bank became more shoal. There were very large islands, and there appeared more to the southward.[6]

He had encountered a serious and complicated obstacle—Torres Strait.

Chapter Eighteen
THE TRAVERSE OF
TORRES STRAIT

15 89

The circumstances described by Torres are important. By *placel*, spelt *praçel* by Prado, Torres meant the submerged bank that extends hundreds of kilometres through Torres Strait, a continental shelf that connects Australia and New Guinea, where the sea ranges in depth from 4 to 5 metres in the north to some 20 metres or less just east of Cape York Peninsula, most of it strewn with reefs and islands. Seeking a passage through, Torres was now forced to steer southwest, passing islands 'without number'. This is a significant statement in attempting to trace the navigator's track. But it does not go far enough. Was 11° South an accurate measurement? Which islands did he pass? And what heading did he take at 11° South, if that figure is correct?

One theory argues that Torres' southing was brief and, turning west, he remained close to the New Guinea coast.[1]

Here the islands referred to by Torres and Prado are matched to those along the big island's shoreline—Daru, Saibai, Dauan and others. Torres' comment that 'we went along on this bank for two months' is interpreted as a coasting, although the few latitudes both he and Prado give are considered irreconcilable with the written descriptions.

Other interpretations run Torres' route deep into the strait with, however, differences. Torres simply wrote, 'we went along this bank for two months, at the end of them we found ourselves in 25 fathoms of depth, and in five degrees of latitude'.[2] Prado comments, 'we went out towards the north and discovered a lofty cape of the great country [New Guinea]'.[3] Neither man identifies clearly the route they took southwest into the strait or the channel through which they finally effected their departure from the *placel* and were able to head north again toward New Guinea. Few dates are given and there are apparent errors. Their voyage through the strait and their exit from it are therefore open to interpretation.

Prado's narrative provides some detail on the islands very briefly mentioned by Torres, and reconstructions of the route have been made by identifying these descriptions with particular islands. Having presumably sailed southwest along the eastern side of the long Warrior Reefs, in the late afternoon of 7 or 8 September the two vessels might have anchored in the lee of what the Spaniards called Isla de los Perros (Island of the Dogs) and located at 10° South. Evidently the population had fled the island, abandoning their dogs, so through the night there was the unnerving sound of continuous and inexplicable howling. In the morning the mariners went ashore, found the deserted village and the dogs, one of which they shot, cooked and ate. It was, said

Prado, 'better than venison'. More of a surprise was a group of women. Prado wrote, 'We selected three of the youngest women and put them on board for the service of the crew'.[4] It appears to be the first stated incident of the kind. The site was possibly the low coral island of Zagai or Dungeness.

Prado also described at some length events on the Isla de Caribes (Island of Cannibals), perhaps today's Yam Island. Anchored near the shore, the *San Pedro* was attacked by several 'very tall men' in outrigger canoes. Gunfire killed one man, whereupon the others dived and disappeared from sight. The weapons found in their canoes were impressive—bows too strong for the Spaniards to bend and extremely heavy, stone-headed clubs a metre long. Prado thought no arquebus-proof helmet could resist a blow from one of these. Landing, the Spaniards found a deserted village 'with numbers of skulls and bones of men they had eaten'.[5]

Other islands which they would necessarily have passed receive little or no mention. An eclipse of the moon was recorded, and on the following night a violent storm enveloped the anchored vessels. It was, Prado wrote:

> as if all the elements had conspired against us; so that at midnight we all made confession and prepared to die; of the two cables with which we were anchored one broke . . . but Jesus was pleased to have mercy on us. So great was the water and sand that entered along the bowsprit that the upper deck of the ship was half blocked up. At dawn the storm ceased . . .[6]

Now, however, they found themselves surrounded by reefs on every side, except in the direction from which they had come and from which the wind was blowing strongly. Torres called a council of his officers, and the solution arrived at was, in Prado's words, 'as if it had come from heaven'. The two vessels remained anchored as the flood tide ran in with the wind. As it ebbed, the men set foresails for steering and weighed, so they rode the retreating tide against the wind. Working in this way for three days, they emerged with the ships unscathed.

Pumice stone, no doubt floated in from some distant volcano, littered the beach of one island. Believing that they stood upon an extinct volcano, the Spaniards named it accordingly, Isla de Vulcan Quemado. This may be Sassie (formerly Long Island). Prado wrote, 'From this we went towards others and reached the largest which greatly resembled the hill of Our Lady of Montserrate'.[7] The island was evidently the rugged 227-metre high Nagheer or Mt Ernest, which reminded Prado of the ravine-riven mountain outside Barcelona, on which stands a Benedictine monastery.

If Prado's Montserrate was indeed Nagheer, Torres, at the masthead of the *San Pedro*, probably scanned his surroundings very thoughtfully. To the west rose the bulky outline of Moa, formerly Banks Island, with Badu unseen behind it, and if the day was clear, the hills of the Prince of Wales island group to the southwest, perhaps the 'very large islands [with] more on the southern part' that Torres later recorded. To the north, northeast and immediately south, reefs and islets rose from the submerged base of the *placel*. Prado simply says that they sailed 'in search of other islands' and mentions a landing where, despite swarms of viciously stinging flies, they obtained a good supply of very clear water. Somewhere among these islands Torres captured 20 people of 'different nations', in order to provide King Philip with first-hand knowledge of the inhabitants of his new domain as well as information on 'other peoples', presumably known to the captives but not to the explorers.

It seems possible that by 3 October 1606 Torres had come far enough south to anchor at the entrance to Endeavour Strait, between Cape York Peninsula and Prince of Wales Island. The argument for this turns upon Torres' reference to 11° South latitude. Reaching this latitude could have carried the ships southwest of the rocky tip of Cape York, which is at 10° 41' South, into Endeavour Strait. It has been argued that this latitude, as given in Torres' letter to the king, was an error. However, in his letter to Quirós, summarised by Quirós in one of his memorials, the 11° South (written by Quirós as '11½°') is repeated, and is therefore unlikely to have been a mistake. When he said 'eleven degrees', Torres evidently meant 'eleven degrees'. There was, of course, no reference to longitude.

Despite these scraps of evidence, the question remains: did Torres actually penetrate far enough south to pass through Endeavour Strait at 10° 50' South, 164 years before James Cook did so, and emerge in this way from the strait and its shallows? We cannot be certain.

The islands of the Torres Strait today constitute part of the Commonwealth of Australia and of the state of Queensland. Torres therefore saw and explored an area of modern Australia. But did he sight the Australian mainland? If the final track of the *San Pedro* and the launch was similar to the one followed by Cook over a century and a half later, then for perhaps two days Torres and his men would have watched as they slowly passed a rocky headland, a carpet of silvery beach and a long stretch of scrub-grown bluffs, without realising that they were at the edge of a continent. The map mentioned by Prado depicted New Guinea as the mainland 'summit' of an antipodal continent. The expedition had just shown that this was not the reality; island-strewn water lay south of New Guinea.

What, then, did Torres believe he was seeing as he followed the shoreline? Another of the 'very large islands'? Interestingly, Prado wrote that on the eve of St Francis' Day, 3 October, 'we found other islands towards the north and among them one bigger than the rest'.[8] Notably the islands were to the north, not the south, the bigger one possibly Prince of Wales Island. What did they see to the south?

Unlike Quirós, Torres was never quick to claim as the sought-after southern continent any new landfall that was not obviously insular. His viewpoint was much more pragmatic. He had looked for Terra Australis Incognita in the Pacific Ocean at 21° South. It had not been there. Did he now doubt that it could be as far north as 11° South? In connection with the map of New Guinea, Prado later remarked that at least at one point, 'we thought it was so', but apparently no one was seriously convinced. It is to be borne in mind that the Spaniards simply did not know where they were. In the world of islands in which they had found themselves, perhaps any land to the south would be thought to be just another island.

An alternative theory considers Torres' reference to 11° South a mistake for 10° South, and suggests that he made his way westward from the strait through one of the more northerly channels lying between the Prince of Wales island group and New Guinea. It was through a narrow, reef-bound channel north of Badu Island that William Bligh sailed in 1792. This accords well with Prado's comment that they passed along 'a very narrow channel', in contrast to Endeavour Strait, which is 11.2 kilometres wide. There was once at least one document that would probably have settled the question of Torres' course. In December 1613, Prado wrote to Philip III through the king's secretary, Antonio de Arostegui, that he was sending to him 'the map of the discovery which Luis Vaez de Torres, Captain of the Almiranta of Pedro Fernández de Quirós, made'.[9] This map has disappeared and so the tantalising question of Torres' route remains.

Whatever Torres' course, the two vessels now emerged from the strait. They had been, according to Prado, 'among these rocks and shoals for 34 days'. Torres says 40 days. Whichever was correct, Prado could now write, 'the wind being a strong breeze we went out toward the north and discovered a lofty cape of the great country [New Guinea], and we steered towards it'.[10] They were in the open waters of the Arafura Sea. The lofty cape was New Guinea's Tanjung Vals, formerly Cape Valsche.

Torres' navigation and ship handling would necessarily have been exceptional. In totally unknown, uncharted waters, he had conned his vessels through a perilous maze of islands, coral reefs, shoal patches and currents for which there was no warning other than a discoloured or breaking sea. During one eight-day period they had fought currents so strong that two men had to be kept at the helm. The onslaughts of storms were further trials survived well under his command.

Perhaps on 8 October, Torres' little squadron rounded the high cape and for three days were out of sight of land, bearing northwards. When land again appeared, they saw mountains lost in cloud, and followed Irian Jaya's coast, low, swampy and deeply green, edged with a network of streams and mud banks.

At the little coastal island of Lakahia the ships anchored and the men went ashore in the boats. They found native houses built high in the trees, 'woven' into the branches, carried off a large pig and gathered oysters and snails, providing themselves with fresh food for several days.

About ten days later they anchored at the mountainous offshore island of Aiduma where a seemingly friendly visit by natives was followed by the approach of a war band in eight large canoes. A volley from the arquebusiers ended the possibility of attack. There were incidents of new interest. Fascinatingly, a huge crocodile surfaced near the ship every evening at seven o'clock. Prado recorded the birth of a child to one of the women taken from the Island of the Dogs. Torres wrote that he found the local people 'better adorned' than those met previously, with weapons that included lengths of bamboo filled with lime, 'with which when fighting they blind their enemies . . . I took possession in Your Majesty's name'.[11] Here Prado completed his fourth and final map of the voyage. Again the map is signed by Prado and credit for the 'discovery' of the area given to Torres. But no one on board knew where they were.

SPE ET METV.

Chapter Nineteen

'WHERE ARE WE?'

15 89

Through the latter part of October and the first days of November the little squadron continued to steer mainly northwest along the steep, much indented coast of southwestern Irian Jaya. Torres, watching the horizon, saw 'an infinity of islands . . . I doubt if in 10 years it would be possible to inspect the coasts of all the islands we saw'.[1] Then there was a joyful surprise. Having gone ashore at a village probably on Sabuda, one of the Pisang Islands, Torres wrote, 'It was here in this land that I found the first iron and bells from China, and other things from there, by which we understood more surely that we were near the Moluccas'.[2] Prado wrote, 'it was a good indication to us to give up the idea that we were lost'. They were on the fringe of the world they knew.

Opposite Sabuda Island the coast Torres was following

opened to a wide expanse of water that today is mapped as Teluk Berau or McClure Gulf. For Torres it presented the question of whether it was a large inlet or a strait separating New Guinea from land to the north that modern maps show as Jazirah Doberai, formerly Vogelkop. In his lost letter to Quirós, Torres seems to have mentioned that New Guinea ended in 1½° South latitude, which would be at Tanjung (Cape) Sele, well to the north. The inlet was then merely a gulf. How did Torres know this? He wrote that they were still in the relatively shallow waters of the *placel* and he had observed 'the great smoothness of the sea'. He continued, 'from what we understand' (*mas de que entendemos*), one side of the opening joined the other 'further back'.[3] It would seem that the Sabuda people had added local knowledge to a seaman's observations.

At three o'clock in the afternoon of 9 November, Torres' vessels anchored between two small, forested islands in the Kepulauan Fam group at the western entrance of Selat Dampier (Dampier Strait). At four o'clock, to the Spaniards' amazement a boat with an awning approached. A man dressed in red emerged from under the awning, hailed the *San Pedro* and in Portuguese asked from what country it had come. Unable to resist a bit of waggery, the Spaniards shouted 'Portugal', but the man retorted promptly: the Portuguese did not call at these islands. The crew had had their fun, and Torres ordered the Spanish ensign run up to the masthead. Still the man demanded where they came from, and when told Spain, came alongside, climbed on board, and embracing the newcomers, welcomed them extravagantly in what Prado called the 'Portuguese fashion'. The Spaniards, he said, were 'valiant cavaliers with swords of gold' and astounding courage.

Prado wrote, 'He asked for wine and drank three times . . . he said that Biliato, his lord, who was governor of two villages, had sent him'.[4] If the strangers were friendly, he would send them provisions. This was welcome news, but the Spaniards had an overriding question: Where were they? They were, the envoy said, at the end of the Papuas, that is, New Guinea, five days' sail from the Moluccan kingdom of Bachan, where there was a Jesuit priest. As well, Don Pedro de Acuña, Governor of the Philippines, had recently seized from the Dutch the clove-rich islands of Ternate and Tidore, which were now occupied by a Spanish garrison. The explorers' joy and relief were overwhelming. Prado wrote:

> It is impossible to exaggerate the pleasure we all felt at such good news and certainly for us he was like an angel, for we already gave ourselves up for lost, and as in the modern maps they colour this great land [New Guinea] as mainland of the summit of the antarctic pole, we thought it was so until this man told us where we were. We gave thanks to God . . .[5]

The land connection of New Guinea to an antarctic continent, which Prado must have seen on a map, had just been proven not to exist.

At ten o'clock the next morning Governor Biliato arrived in a large, galley-like native craft, with an accompanying boat loaded with pheasants, kids, cocks and hens, as well as birds of paradise and brilliantly coloured parrots. With the expedition's trade goods gone with the flagship, Torres' men bartered as best they could with old clothes and shoes and knives they did not need. For Biliato, however, a length of green taffeta was found, and he gave in return a handsome 10-year-old boy.

Surprisingly, Biliato was accompanied by a Moor marked on the cheek with the letter S and the image of a nail, *clavo* in Spanish, the combination creating the word *esclavo* (slave). The man spoke good Italian and described how as a boy he had been captured by Spanish forces under Don Juan of Austria, possibly when subduing Moorish descendants in Granada, and sold in Seville to merchants from Manila. In Manila he and others had stolen a boat and escaped, sailing through the Philippine archipelago to the Moluccas.

Prado and Torres questioned him on the resources of the country, and learned of gold, pearls, sandalwood, black pepper and red gemstones. There were also swine and buffalo, ginger and coconuts. The Chinese, the Moor said, had come to barter for gold and black pepper, but had not appeared for the past three years. They also would have questioned the Moor at length on the passage between the Moluccas and Manila, which he had sailed during his escape. Prado referred to this knowledgeable man as Alfaqui, 'the learned one'.

Biliato's departure was saluted with the firing of three cannon. Two days later the Spaniards sailed, steering southwest across the Halmahera Sea. They were now well into the Moluccas, a region very different from the New Guinea coast. Torres wrote:

> we found Moors [Moslems] clothed, having artillery for service, such as falconets and culverins, muskets and side arms; they carry on conquests of the people they called the Papuas, and preach to them the sect of Mahomet; these Moors traded with us . . . [and] gave news of the events at the Moluccas, and of Dutch ships . . . in all this land there is much gold and other things, good spices such as pepper and nutmeg.[6]

About 16 November the southeast tradewinds on which the ships had sailed so long slowly died, and the northwest monsoon broke upon them. For eight days the *San Pedro* and the launch tacked, fighting contrary currents to reach the island of Pulao Obi (Obi Major). Torres now turned north, the *San Pedro* and the launch making their way through a deep clear channel to anchor at four in the afternoon near Pulao Bacan, known to the Spaniards as the island of Bachan. Fifteen minutes later 'a covered canoe appeared from there, with a man clothed in red'.[7] The Spanish crew could not resist their joke of claiming again to be Portuguese, but their visitor's response was scornful: 'You are nothing but Dutchmen.' This could not be tolerated, and the Spanish flag rose quickly at the poop.

The man, who was Portuguese, assured himself of the Spaniards' nationality, climbed aboard and as Prado wrote, 'embraced us with much affection and satisfaction'. A Dutch ship, he explained, had left just eight days before, with 24 pieces of artillery but only ten very sick men on board. As to the Portuguese, he was the 'servant of the King of Bachan, lord of this country', who wished to honour the Spaniards by coming on board.

The king arrived just at nightfall, requesting that no guns be fired, as they terrified him. An old carpet had been retrieved from somewhere in the ship and on this, spread out in the gallery of the poop, the king seated himself. Dressed in purple taffeta but bareheaded and barefooted, he had with him twelve guardsmen armed with 'butcher's' knives, one bearing a naked sword and round shield, the insignia of kingship. An attendant carried in one hand a cone-shaped silver holder inserted with tobacco pipes and in the other a silver flask of native wine. Present also was the king's auditor. The Portuguese acted as interpreter.

The king enquired after his brother, the King of Spain: his health and his age, the number of his children, and whether he was a great sportsman with the round net for catching sardines and other fish. Prado informed him that Philip was a fisherman of kingdoms and enumerated his various domains. The king then asked for Spanish wine, drank 'three times', smoked and went to sleep on the carpet. With armed strangers on board, Torres stationed guards throughout the ship.

The next day, at the suggestion of the king, the *San Pedro* and the launch were towed by native craft to a better anchorage, where they arrived on 16 December, a year less six days from the date of their departure from Peru. A soldier was sent the 24 leagues in a native boat to Ternate to advise Juan de Esquibel, the Spanish campmaster, of their arrival. Christmas was spent at the anchorage, with the company of the Jesuit, Father Fonseca.

The king of Bachan had asked Torres for help in subduing some rebels, people that had been placed under his jurisdiction by the Spanish governor, Pedro de Acuña, when he took Ternate and some of the surrounding islands. Torres' messenger cleared this with Juan de Esquibel, and after Christmas Torres prepared for the attack. He wrote to Philip III:

> I decided on forty Spaniards and four hundred Moors of the King of Bachan, then I made war against them and in only four hours I put them to flight and took the fort, and put the King of Bachan in possession of it in Your Majesty's name: to whom I administered the usual oaths stipulating with him that he should never go against Christians and should always be a faithful vassal to Your Majesty.[8]

Prado described a clandestine approach by wading through a mangrove swamp and the charge at the gate of the square, moated stone fort, the shouting of 'Santiago!', the drum roll and volleys from the arquebuses, so that the defenders fled 'pell mell' through a side door. Shortly after the little fleet was again under sail, threading its way northward among the islands that edge the west coast of Halmahera.

SPE ET METV.

Chapter Twenty
TERNATE AND MANILA

15 89

Some 23 kilometres off the west coast of the island of Halmahera two small, round islands, dominated by their volcanos, rise out of the Molucca Sea. In size no more than about 116 square kilometres each, they were for centuries the centre of fierce rivalries, their respective sultans, who were traditional enemies, the Portuguese, the Dutch and the Spanish all competing for the priceless clove trade. Briefly there was an English element as well, when in 1579 Francis Drake arrived at Ternate, loaded a cargo of six tons of cloves and allegedly reached a verbal agreement with the sultan on a future English trading post.

A treaty between the sultan of Ternate and the Dutch evidently galvanised the Spanish into action. In February 1606 Pedro Bravo de Acuña, Governor and Captain-General of the

Philippines, sailed from Manila with a large fleet of galleons, galleys, barges and other ships and boats carrying over 1000 Spanish troops, 400 Filipino soldiers and artillery, munitions and supplies for nine months. Tidore was taken without opposition and on 1 April the attack was launched on Ternate. The extensive fortifications were stormed, the sultan and the Dutch escaped by boat and the Spaniards entered a deserted fort. Three small fortlets were quickly occupied. Quantities of cloves, cloth, 2000 ducats and Portuguese and Dutch artillery and munitions were taken.

Within days the sultan had returned to swear homage to King Philip of Spain, as did the Sultan of Tidore and the King of Bachan, while lesser chiefs promised allegiance to these rulers. Acuña extracted promises that restricted the clove trade to the Spanish, remitted to the local people some of the tribute they had paid to the sultans, and saw work begun on improved defences for both Ternate and Tidore. The invasion fleet then departed with Ternate's sultan as a hostage. Six hundred Spanish troops were left on the two islands under Juan de Esquibel. Tidore had remained peaceful but there had been insurgency on Ternate and some of the other islands. Within two months of Acuña's departure a cutter arrived from Manila with the news that he had suddenly died.

Probably on 6 January 1607, Torres and his vessels reached Ternate. Gun salutes to the first Spanish community they had seen in more than a year brought out a pilot, who guided them to a sheltered anchorage at the northern end of the island. The campmaster Esquibel welcomed them with enthusiasm. A relief ship was expected from Manila in three or four months, at the end of a year's occupation, and Esquibel wanted Torres and his men to remain until then. Torres noted that of Esquibel's garrison half the people 'were no more', but whether through sickness or other difficulties he did not say. Torres saw advantages in waiting. His ships needed attention and his men rest, and speaking to the pilots of the relief ship, who would have just made the voyage from Manila, could be very useful. Possibly he also awaited a monsoonal change.

Torres and his command remained on the island for three months. Their presence bolstered the security of the Spanish garrison and the opportunity to associate with a different group of men was enjoyed by everyone. Two reports written by Esquibel at this time further confirm Torres' top position in the little squadron, referring to him as *capitan y cabo* (captain and commander).

In late April a ship arrived from Manila with supplies and reinforcements. Torres and his pilot Fontedueña would have conferred at length with the captain and pilot of the new arrival on the recommended route through more than 1600 kilometres of island-scattered ocean. On 1 May the *San Pedro* and her company weighed and steered out of the Ternate anchorage. Torres left with Esquibel the launch and 20 men, as was 'expedient' for King Philip's service. There is no record of the feelings of the men thus left at this distant outpost of empire.

Alone the *San Pedro* traversed the Celebes Sea, past the little islands that are today the northernmost extensions of Indonesia. At the southern Philippine islands of Sarangani the ship swung west to the island of Jolo, and then north to the big, sprawling island of Mindanao and the native settlement called by Prado Rio de Canela (Cinnamon River), which in time would become the city of Zamboanga. Here Torres traded for a large cargo of cinnamon. He then continued north, running along the archipelago's central islands—Negros, Panay, Mindoro. The weather seems to have been fair, a blue ocean with the pale blue outlines of land to the east under a sunfilled sky.

On 22 May 1607, the *San Pedro* entered Manila Bay and slowly made its way into the inner harbour of Cavite, where the vessel was boarded by officials. Since the death of Acuña the previous December the Philippines had been administered by a governing council, the *audiencia*, headed by an acting governor, Rodrigo de Vivero.

The Spanish city of Manila occupied a rough triangle of land wedged between the bay and the Pasig River, encircled by a wall of hewn stone and moated on its landward sides. There were straight cobblestone streets with stone kerbs and footpaths of granite brought in as ships' ballast, residences mainly of stone or brick, with entrance doors just off the footpath and upstairs balconies enclosed by ornamental iron grilles. The city had been all but razed by earthquake and fire in 1603 and new regulations required stone and tile construction, instead of the flammable bamboo and nipa thatch of earlier years. Similar in plan to Spain's colonial cities in the Americas, its fortress of Santiago occupied the apex of the triangle, overlooking the junction of the river and the bay. There were handsome administrative buildings, hospitals, the viceregal palace and the cathedral, and other churches and religious houses. Outside the walls and across the river spread the native villages and the large Chinese settlement.

Nine days after Torres' arrival in Manila he learned that the *San Pedro y San Pablo* had reached Mexico with Quirós safely on board. Correctly, he assembled a report to his former commander on the latter part of his voyage, dating the finished document 15 June 1607. The letter has apparently been lost, but a somewhat confused summary of its contents exists as part of one of Quirós' lengthy memorials to Philip III. The summary is incomplete, for Quirós evidently omitted information detrimental to his own argument for a new expedition or to his insistence that Espíritu Santo was part of Terra Australis Incognita. Nor did he mention that Torres had found no continent at 21° South latitude, or that the newly discovered coastline was the southern littoral of New Guinea.

Torres' orders to now refit and reprovision his ship for the voyage to Spain met with a wall of obstruction. He was told that the necessary stores could not be spared and subsequently that his ship and its crew were needed for local service. Even with the arrival from Mexico of two ships with soldiers, munitions and money, his persistent requests were denied.

There is no ready explanation for this conduct on the part of the *audiencia*, but there are partial answers. The islands of the Philippines were by no means fully subdued. Settlements in northern Luzon and on other islands, where missionaries sought to organise communities and Christianise the people, were frequently under attack by natives. A list of Dominican friars gives the cause of death in many cases as simply *flechado*, 'arrowed'. Soldiers and money for their support were thinly spread.

A personal element may well have entered the picture too. The influential chief magistrate of the port of Cavite, Felipe Corzo, had been captain of the galeot under Quirós during Mendaña's disastrous voyage of 1595. For reasons now not clear, Corzo hated Quirós. The arrival of the *San Pedro* at Cavite seems to have reawakened this enmity, for on 15 July, less than two months after Torres' arrival, Corzo wrote to Philip III, urging that Quirós should not be entrusted with any new expedition. One can only assume that Corzo's anger extended to anyone associated with Quirós or who perhaps might represent Quirós more favourably in Madrid. Probably unfortunate for Torres was the recent death of the energetic and enterprising governor, Pedro Bravo de Acuña. The decisions of this vigorous man may have been very different from those of a cautious *audiencia*.

Whatever the reason behind the decisions to refuse assistance and then take his ship, unsurprisingly Torres was very angry. On 12 July 1607 he wrote to King Philip. Without heading or salutation, the letter began boldly:

> Being in the city of Manila at the end of a year and a half of navigation and discovery among the lands and seas in the unknown southern part, and seeing that in this Real Audiencia of Manila they have not hitherto thought fit to give me dispatch for completing the voyage Your Majesty commanded . . . I have thought proper to send a person to give account to Your Majesty . . .[1]

The person chosen 'to give account' to the king was Fray Juan de Merlo, one of the Franciscans who had accompanied Torres on the *San Pedro*, and it was to him that Torres entrusted his letter for delivery. As events transpired, Merlo did not travel beyond Mexico, but Torres' letter arrived in Madrid, presumably by official mail, on 22 June 1608, together with five maps signed by Prado and four unsigned and undated watercolour drawings of island natives. Only four of the maps, each of an anchorage location, survive. If the fifth was a track chart of the voyage or part of it, it is a very regrettable loss. The origin of the watercolours is unknown, but probably they are meant to represent the inhabitants of the four anchorage sites mapped by Prado. If so, there are many errors in the detailing of the scenes. Torres made no mention of either maps or drawings.

The letter consisted of just nine pages summarising the 17-month journey, during which, as he noted, he had lost just one Spaniard. He then repeated his complaint of ill-treatment in Manila and concluded:

> [I] know not when I shall be able to set out from here to give account to Your Majesty, whom may Our Lord protect and prosper for sovereign of the world. Done in Manila, on the 12th of July, in the year 1607: Your Majesty's Servant, Luis Baes de Torres.[2]

The letter was very much as Luis Váes de Torres himself comes across the years: confident and courageous, respectful without excess even to his king, factual and to the point in his thinking.

Nearly a year later, on 6 June 1608, Torres affixed his signature to a certificate testifying to the accuracy of the information presented in Prado's completed *relación*. This is the final piece of evidence we have of Luis Váes de Torres. The brave and competent leader, the skilled navigator of treacherous, uncharted waters, the European discoverer of the strait separating New Guinea from the Australian continent, simply vanishes from history. Speculation can range from an unrecorded departure from Manila and loss at sea to death from disease or

in a tavern brawl. But no thread of evidence exists for any of
this. Of Torres' twenty captives meant for the king's inspection
there is also no record.

Prado appears to have remained in Manila until 1610. By
December 1613 he was in Goa. In letters written on the 24th
and 25th he noted his plans to return to Spain by way of Ormuz,
Aleppo and Venice. The date of his arrival in Spain is unknown,
but at some point afterwards he became, as he wrote, 'a monk of
our father Saint Basil the Great of Madrid'.[3] Now he evidently
sat down to re-write his *relación*. The result was a holograph
copy of the original written in Manila, with additions relating
to more recent events, and the accompanying certificate signed
by Torres and four others, copied in Prado's hand. The original
relación and certificate have never been found. What was found
in a collection of documents sold at auction in London and
published in 1930 was the later, rewritten copy. That Prado could
have altered certain events in the account to please himself is, of
course, obvious.

SPE ET
METV.

Chapter Twenty-one

A LEGACY OF ISLANDS

15·89·

T he three voyages of 1567–1607 were Spain's final major attempts at the exploration of the South Pacific Ocean. Considerable expenditures had yielded none of the expected results in new sources of wealth, converts to Catholicism or colonies. Most importantly, the expense had become an unwelcome addition to what for Spain was an increasingly troubled economy.

The Spanish Empire was territorially, militarily and economically an intrinsic part of Europe. The European policies of her monarchs Charles I, who was also Charles V of the Holy Roman Empire, and Philip II incurred almost ceaseless wars against France, the Papacy, Germany's Protestant princes, England and Spain's own rebellious subject states. In addition there were campaigns to counter the onslaughts of the Ottoman Turks.

The cost of these wars was enormous and Spain was not itself a wealthy nation.

In the later 16th century the pressure on Spanish finances was unrelenting. In 1552 Charles' German bankers refused to make him further loans. Philip II defaulted three times on his loans with the banking house of Fugger despite the bullion revenues that became a steady money transfusion from the Americas, especially after 1550. These, however, had frequently been assigned to the king's creditors before the funds had even left the New World. Under such circumstances profitless Pacific exploration was one liability that could be avoided. No further expeditions were despatched in search of Terra Australis Incognita.

What then was accomplished by Spain's final voyages of exploration into the South Pacific? The contribution to knowledge made by Mendaña, Quirós and Torres and the men who sailed with them was won through desperate effort and fierce courage, often with a result they had not sought. Yet three major island groups, the Marquesas Islands, the Solomon Islands and Vanuatu's Espíritu Santo were added to the sum of European geographical knowledge, gradually appearing on 17th century maps, although often inaccurately placed and imaginatively drawn. Countless small, individual islands were noted, but with locations inexactly recorded in terms of latitude and particularly of longitude, many were 'lost' until 'rediscovered' by later navigators. The narratives of some of the voyagers included lengthy descriptions of people, island life, terrain, plants and animals that enhanced European awareness of another part of the world. Prado described the marsupial cuscus of New Guinea. Others provided details on the construction of island boats. The four pictures of native groups sent to the Spanish king by Torres were no doubt among the first illustrations of the kind seen at a European court.

Perhaps the most important result of the three voyages, the traverse of Torres Strait, became known to the world only in fragments. For fear that Spain's enemies would profit by knowledge of this discovery, Torres' letter to Philip III was filed in the archives at Simancas, unseen until brought to light by the Spanish historian Muñoz in 1782, twelve years after James Cook had rediscovered the route. Prado's narrative, similarly, was lost until 1930.

Some of the facts, however, did emerge. During their time in Manila, Torres and Prado would have passed on to others something about their voyage. Quirós' son Francisco and his nephew Lucas evidently had some of that navigator's papers. In 1621 the Spanish historian Hernando de los Ríos Coronel mentioned having read Prado's account, and in about 1630–33 Juan Luis Arias de Loyola, who had known Quirós, sent a memorial to King Philip IV which included a short reference to Torres' voyage, details that presumably came from the explorer's letter to Quirós.

The Arias document acquired an interesting connection with James Cook's 1770 journey up the Australian east coast. In about 1765 the English hydrographer Alexander Dalrymple had found the Arias memorial and drawn a chart showing Torres' presumed track along the New Guinea south coast. The chart and Dalrymple's book on South Pacific discoveries, including Arias' comments, were given to Joseph Banks for his voyage with Cook in 1768, and undoubtedly suggested to Cook the possibility of a strait.

Of charts stemming directly from the voyages few have

survived. An original map mentioned by Hernando Gallego has been lost, but copies seem to exist, as well as a rutter issued by the Casa de Contratación to official pilots, on which the Solomon Islands appeared. Quirós, a skilled cartographer, displayed charts of the Pacific in Rome and at the Spanish court, and later claimed to have drawn and made over 200 maps and globes. Of these only one is extant, dated 1598 and initialled P.F.Q. Prado's four surviving maps are of anchorage sites; four others mentioned are now missing.[1]

Information from these works, however, may have found its way onto the charts of Manuel Godinho de Erédia (1563–1622) and from these onto maps of New Guinea, dated from the 1620s to the 1640s, by several Portuguese cartographers. An Erédia chart of the South Pacific may be the earliest map showing the island discoveries of Quirós and Torres' exploration along the south coast of New Guinea. A map by the great Dutch cartographer Hessel Gerritsz, dated 1622 and amended to 1634, includes a legend that refers to Torres' journey—mistakenly ascribed to Quirós—westward on 10° South for 40 days. The Pacific findings of Mendaña and Quirós also appear, taken, according to Gerritsz, from charts he saw in Seville in 1618.

Thus the findings of Spain's last Pacific Ocean explorers were added to the expanding image of the world, early links in the chain of exploration carried forward by their successors, among them Roggeveen, Wallis, Carteret, Bougainville, La Pérouse, Malaspina and Cook.

A literary dimension also has added to general knowledge of the Mendaña-Quirós-Torres explorations. In the early 20th century the relative documents were published both in Spanish and in English and in other European languages as well. A thorough examination of Torres' voyage through Torres Strait was published in 1980. A number of journal articles have also been written. In recent years an epic poem and an imaginative novel, both on Quirós, have been published.

As Europeans became aware of the actual continent of Australia, theoretical connections developed between it and the Spanish island discoveries. A map of 1753 by Jacques Nicolas Bellin shows Australia joined to a long finger-like New Guinea and Espíritu Santo as a part of what is now northern Queensland. Although conclusively disproved by Cook in 1770, this cartographic error fostered the idea among some that Quirós had 'discovered' Australia, a notion revived as late as the end of the nineteenth century.[2]

Mistaken as it was, Quirós' vision of a chimerical antarctic continent was part of the beliefs that spurred other exploration. On his 1772–75 voyage into the Pacific, when James Cook crossed the Antarctic Circle to the rim of Antarctica, he saw no lush welcoming continent, only ice. The mirage was finally dispelled.

GLOSSARY

ADELANTADO—governor of a frontier region; expedition commander

AGUACIL MAYOR—high constable; head of law enforcement

ALMIRANTA—by custom the term for the second vessel of an expedition, usually captained by the expedition's second-in-command with the title of admiral

ARQUEBUS—a smooth-bore matchlock gun with a stock resembling that of a rifle

ARQUEBUSIER—soldier armed with an arquebus

AUDIENCIA—the governing body of a province, composed of a president and four judges

BAJOS—a low area; lowland; shoals

BEAT—to sail as closely as possible to the wind by alternating tacks

BINNACLE—a stand with a housing for the compass, compensating magnets and lights by which the compass can be read at night

BLUNDERBUSS—a short musket with an expanded muzzle to scatter shot or bullets at close range

BONNET—an additional piece of canvas laced to a sail to create a larger sail area

BRIGANTINE—at this period usually an undecked boat of 4 to 5 tonnes, designed to carry sailors and soldiers for coastal exploration

CAPITANA—by custom the term for an expedition's flagship; the commander of the expedition referred to as the general

CAULKING—filling a ship's seams with oakum or cotton

CÉDULA—a royal order; certified copy of a document

CHANNELS—ledges built out on the side of a ship to increase the spread of the shrouds

CLARION—a small-bore trumpet

CORSAIR—a pirate or privateer

COURSES, COURSE—sails set to the lower yards, e.g. fore course or main course

CULVERIN—a type of 16th–17th century cannon

FALCONET—a light 16th century cannon

GALEOT—a vessel combining sails and banks of oars; related to the galley

GALLERY—a balcony-like platform along the stern of a ship

HALBERD—a weapon with a long shaft and an axe-like cutting blade

LEAGUE—unit of measure roughly equal to 3 miles or 5 kilometres

LICENCIADO—licenciate; holder of a degree in a Spanish university

MARAVEDÍ—a small value coin

NAO—sturdy broad-beamed and three-masted ship built to accommodate large cargoes; later versions referred to as galleons

NIPA—a palm of southeast Asia whose foliage is much used for thatching

OAKUM—strands of old rope used for caulking

'ONBRE A LA MAR'—man overboard

PAY A SHIP'S SEAMS—to seal a ship's seams with pitch after they have been filled with oakum or cotton

PESO CORRIENTE—monetary unit valued at about 300 maravedís

POBLADOR—founder of a community

RELACIÓN—a narrative, report or memoir

ROAD, ROADSTEAD—an anchorage that is some distance from shore

RUTTER—a set of navigational instructions for finding a course at sea

SALVE REGINA—a prayer to the Virgin Mary

SHROUDS—lines supporting the mast

STAND OFF AND ON—to sail towards and then away in order to maintain position, usually at night or while waiting

STRIKE—to lower, e.g. strike the topmast

TOP—a platform surrounding the head of a lower mast used to spread the rigging and serve as a platform; lookouts stood here (masts were in upper and lower sections)

WEIGH—to raise anchor

WHIPSTAFF—an extended lever fitted to the rudder for steering a ship

NOTES

Prologue

1 Celsus Kelly, O.F.M. (trans. and ed.), *La Austrialia del Espíritu Santo: The Journal of Fray Martín de Munilla O.F.M. and other documents relating to the voyage of Pedro Fernández de Quirós to the South Sea (1605–1606) and the Franciscan Missionary Plan (1617–1627)*, vol. I, Hakluyt Society, Cambridge, 1966, p. 137.

2 Clements Markham (trans. and ed.), *The Voyages of Pedro Fernandez de Quiros, 1595 to 1606*, vol. I, Hakluyt Society, Nendeln/Liechtenstein, 1967, pp. 180–201.

Chapter 1 Gold, Souls and a Mythical Continent

1 Aristotle, *Of the Heavenly Bodies*, Book II, *The Works of Aristotle*, vol. I, Encyclopaedia Britannica, Inc., University of Chicago, Chicago, 1986, p. 388.

2 T.M. Perry, *The Discovery of Australia*, Nelson, Melbourne, 1982, p. 20.

3 Celsus Kelly, O.F.M. (trans. and ed.), *La Austrialia del Espíritu Santo: The Journal of Fray Martín de Munilla O.F.M. and other documents relating to the voyage of Pedro Fernández de Quirós to the South Sea (1605–1606) and the Franciscan Missionary Plan (1617–1627)*, vol. I, Hakluyt Society, Cambridge, 1966, p. 94.

4 Juan Luis Arias de Loyola, 'Memorial to Philip III' in R.H. Major (ed.), *Early Voyages to Terra Australis, now called Australia*, Hakluyt, London, 1859, p. 15.

5 ibid., p. 16.

6 Cornelius de Jode 1593, 'Novae Gvineae Forma & Situ', *Speculum Orbis Terrae*, in Henry N. Stevens, *New Light on the Discovery of Australia as Revealed by the Journal of Captain Don Diego de Prado y Tovar*, Hakluyt Society, London, 1930; Kraus Reprint, Nendeln/Liechtenstein, 1967, p. 18.

CHAPTER 2 THE FIRST VOYAGE: 1567–1569

1 The seagoing ability of such a raft was demonstrated in 1947 by the adventurer Thor Heyerdahl, sailing in one from South America to Polynesia.

2 'Galapagos Islands', in *The New Encyclopaedia Britannica*, Encyclopaedia Britannica, Inc., Chicago, vol. 5, p. 80.

3 Celsus Kelly, *Calendar of Documents: Spanish Voyages in the South Pacific from Alvaro de Mendaña to Alejandro Malaspina 1567–1794 and the Franciscan Missionary Plan for the Peoples of the Austral Lands 1617–1634*, Franciscan Historical Studies in association with Archivo Ibero-Americano, Madrid, 1965, p. 93.

4 Several years later Juan de Iturbe, overseer and comptroller of the expedition of 1605–06, referred to the *capitana* as *Los*

Tres Reyes and stated that on her stern she bore the inscription '*Los Reyes es nombre mío, porque sea guía mía la estrella que fué su guía*' (The [Three] Kings is my name, that the star which guided them be also my guide); in Celsus Kelly, O.F.M. (trans. and ed.), *La Austrialia del Espíritu Santo: The Journal of Fray Martín de Munilla O.F.M. and other documents relating to the voyage of Pedro Fernández de Quirós to the South Sea (1605–1606) and the Franciscan Missionary Plan (1617–1627)*, vol. II, Hakluyt Society, Cambridge, 1966, p. 273. No such inscription is mentioned in the contemporary record.

5 Kelly, *Calendar of Documents*, p. 96.

6 The Governor of Peru to the Council of the Indies, 2 April 1567, in Kelly, *Calendar of Documents*, p. 96.

7 William Amhurst Thyssen-Amherst, Baron Amherst, and Basil Thomson (trans. and eds), *The Discovery of the Solomon Islands by Alvaro de Mendaña in 1568,* Hakluyt Society, Nendeln/Liechtenstein, 1901, 1967, vol. I, p. 188.

8 Dates given for specific events vary in the different narratives. Catoira's, however, usually matched with the holy days he names, and can be checked.

9 Martin Fernández de Navarrete, *Colección de los viajes y descubrimientos que hicieron por mar los Españoles desde fines del siglo XV*, Tomo IV, in Walter Brownlee, *The First Ships Round the World*, Cambridge University Press, Cambridge, 1974, p. 18.

10 Celsus Kelly (ed.), *Austrialia Franciscana*, vol. III, *Documentos sobre la expedición de Alvaro de Mendaña a las Islas de Salomón en el Mar del Sur (1567–1569),* Franciscan Historical Studies in collaboration with Archivo Ibero-Americano, Madrid, 1967, Introduction, pp. xxii–xxiii.

11 Celsus Kelly (ed.), *Austrialia Franciscana*, vol. IV, *Documentos sobre la expedición de Alvaro de Mendaña a las Islas de Salomón en el Mar del Sur (1567–1569)*, 'Relaciones de la expedición y una selección de documentos posteriores sobre la misma', Franciscan Historical Studies in collaboration with Archivo Ibero–Americano, Madrid, 1967, Introduction, p. xvi.

12 O.H.K. Spate, *The Pacific Since Magellan,* vol. 1, *The Spanish Lake,* ANU Press, Canberra, 1979, p. 121.

13 Celsus Kelly (ed.), *Austrialia Franciscana*, vol. II, *Relaciones de Alvaro de Mendaña al Rey Don Felipe II y de Gómez Hernández Catoira al Gobernador del Perú, Don Lope García de Castro, sobre la expedición de Mendaña a las Islas de Salomón en el Mar del Sur (1567–1569)*, Franciscan Historical Studies (Australia) en colaboración con Archivo Ibero-Americano (Madrid), Madrid, 1965, p. 33.

14 Amherst and Thomson, *The Discovery of the Solomon Islands,* vol. II, p. 220.

15 ibid., vol. I, p. 99.

16 ibid. Méndez had been one of the two sailors who swam out to the boy who had fallen overboard a few days earlier.

17 ibid., p. 12.

Chapter 3 Las Yslas de Salomón

1 'a punto de guerra', *Relación*, Alvaro Mendaña de Neira, in Celsus Kelly (ed.), *Austrialia Franciscana*, vol. III, *Documentos sobre la expedición de Alvaro de Mendaña a las Islas de Salomón en el Mar del Sur (1567–1569)*, Franciscan Historical Studies in Archivo Ibero-Americano, Madrid, 1967, p. 194.

2 William Amhurst Thyssen-Amherst, Baron Amherst, and Basil Thomson (trans. and eds), *The Discovery of the Solomon*

Islands by Alvaro de Mendaña in 1568, Hakluyt Society, Nendeln/Liechtenstein, 1901, 1967, vol. II, p. 230.

3 O.H.K. Spate, *The Pacific Since Magellan,* vol. 1, *The Spanish Lake*, Australian National University Press, Canberra, 1929, p. 123.

4 Amherst and Thomson, *The Discovery of the Solomon Islands* vol. I, p. 174.

5 ibid., p. 175.

6 ibid., p. 170.

7 ibid., vol. II, pp. 237–8.

8 Celsus Kelly, *Calendar of Documents: Spanish Voyages in the South Pacific from Alvaro de Mendaña to Alejandro Malaspina 1567–1794 and the Franciscan Missionary Plan for the Peoples of the Austral Lands 1617–1634,* Franciscan Historical Studies in association with Archivo Ibero-Americano, Madrid, 1965, Introduction, p. 30.

9 Amherst and Thomson, *The Discovery of the Solomon Islands*, vol. I, Note 2, p. 172.

10 ibid., vol. II, p. 254.

11 'es gente de gran fuerça i los arojan con gran furia', in 'Relaciones de Hernan Gallego', in Kelly, *Austrialia Franciscana*, vol. III, p. 81.

12 Amherst and Thomson, *The Discovery of the Solomon Islands*, vol. II, p. 201.

13 ibid., vol. I, p. 136.

CHAPTER 4 GUADALCANAL

1 'excede a todas en ser grande y en tener buena apariençia, que cierto no se puede senificar', 'Relación de Gómez Hernández de Catoira', in Celsus Kelly (ed.), *Austrialia Franciscana*, vol. II, *Documentos sobre la expedición de Alvaro de*

Mendaña a las Islas de Salomón en el Mar del sur (1567–1569), Franciscan Historical Studies en colaboración con Archivo Ibero-Americano (Madrid), Madrid, 1967, p. 96.

2 William Amhurst Thyssen-Amherst, Baron Amherst, and Basil Thomson (trans. and eds), *The Discovery of the Solomon Islands by Alvaro de Mendaña in 1568*, Hakluyt Society, Nendeln/Liechtenstein, 1901, 1967, vol. II, p. 310.

3 ibid., p. 374.

4 ibid.

5 ibid., p. 387.

6 ibid., vol. I, pp. 56–7.

7 ibid.

CHAPTER 5 THE LAST ANCHORAGE

1 William Amhurst Thyssen-Amherst, Baron Amherst, and Basil Thomson (trans. and eds), *The Discovery of the Solomon Islands by Alvaro de Mendaña in 1568*, Hakluyt Society, Nendeln/Liechtenstein, 1901, 1967, vol. II, p. 401.

2 ibid., vol. I, p. 180.

3 ibid., vol. II, pp. 408–9.

4 ibid., vol. I, p. 184.

5 ibid., p. 180.

6 ibid., vol. II, p. 426.

CHAPTER 6 THE JOURNEY BACK

1 William Amhurst Thyssen-Amherst, Baron Amherst, and Basil Thomson (trans. and eds), *The Discovery of the Solomon Islands by Alvaro de Mendaña in 1568*, Hakluyt Society, Nendeln/Liechtenstein, 1901, 1967, vol. II, p. 445.

2 ibid., vol. I, p. 73.

3 In quotation from *Relación Breve* in Celsus Kelly, *Calendar of Documents: Spanish Voyages in the South Pacific from Alvaro de Mendaña to Alejandro Malaspina 1567–1794 and the Franciscan Missionary Plans for the Peoples of the Austral Lands 1617–1634*, Franciscan Historical Studies in association with Archivo Ibero-Americano, Madrid, 1965, p. 109.

4 Celsus Kelly (trans. and ed.), *Austrialia Franciscana*, vol. II, *Relaciones de Alvaro de Mendaña al Rey Don Felipe II y de Gómez Hernández Catoira al Gobernador del Perú, Don Lope García de Castro, sobre la expedición de Mendaña a las Islas de Salomón en el Mar del Sur (1567–1569)*, Franciscan Historical Studies (Australia) en colaboración con Archivo Ibero-Americano, Madrid, 1965, pp. 217–8.

5 Amherst and Thomson, *The Discovery of the Solomon Islands*, vol. I, p. 180.

6 Colin Jack-Hinton, *The Search for the Islands of Solomon, 1567–1838*, Clarendon Press, Oxford, 1969, p. 80.

7 Pedro Sarmiento de Gamboa, Preface, *History of the Incas*, in Jack-Hinton, *The Search for the Islands*, op cit, 1969, p. 8.

8 Amherst and Thomson, *The Discovery of the Solomon Islands*, vol. I, p. B2.

CHAPTER 7 THE SECOND VOYAGE: 1595–1596

1 William Amhurst Thyssen-Amherst, Baron Amherst, and Basil Thomson (trans. and eds), *The Discovery of the Solomon Islands by Alvaro de Mendaña in 1568*, Hakluyt Society, Nendeln/Liechtenstein, 1901, 1967, vol. I, p. 181.

2 Celsus Kelly, *Calendar of Documents: Spanish Voyages in the South Pacific from Alvaro de Mendaña to Alejandro Malaspina 1567–1794 and the Franciscan Missionary Plans for the Peoples*

of the Austral Lands 1617–1634, Franciscan Historical Studies in association with Archivo Ibero-Americano, Madrid, 1965, p. 123.

3 Amherst and Thomson, *The Discovery of the Solomon Islands,* vol. I, Introduction, p. lxvii.

4 Celsus Kelly (ed.), *Austrialia Franciscana,* vol.V, *Una selección de noventa documentos sobre las negociaciones de Alvaro de Mendaña en la Corte de Madrid, en Panamá y el Perú para poblar las Islas de Salomón (1572–1595),* Franciscan Historical Studies (Australia) in collaboration with Archivo Ibero-Americano, Madrid, 1967, Introduction, p. 6.

5 There were several *pesos* of different value. The *peso corriente* was equal to about 300 *maravedís*; the *peso de oro* (gold) ranged in value from 450 to 576 *maravedís*. The *maravedí* was worth two *blancas,* a copper coin of the lowest value.

6 Clements Markham (trans. and ed.), *The Voyages of Pedro Fernández de Quirós, 1595–1606,* Hakluyt Society, Cambridge, 1904, 1967, vol. I, p. 8.

7 ibid., p. 9.

8 Common translation for the Spanish or Spanish American peseta or peso.

9 Gerard Bushell (ed.), en colaboración con Celsus Kelly, *Austrialia Franciscana,* vol.VI, *Documentos sobre la expedición de Alvaro de Mendaña para poblar de las Islas de Salomón (1595–1597), Relaciones de la población de las Islas de Santa Cruz, el fracaso y sus consecuencías inmediatas,* Franciscan Historical Studies in collaboration with Archivo Ibero-Americano, Madrid, 1973, Introduction, p. xvi.

10 Markham, *The Voyages of Pedro Fernández de Quirós,* vol. I, pp. 13–14.

Chapter 8 '. . . we opened fire on them'

1 Pedro Fernández de Quirós, 'An account of the journey made by the adelantado Alvaro de Mendaña de Neira for the discovery of the Solomon Islands', in Antonio de Morga, *Successos de las Islas Filipinas*, J.S. Cummins (trans. and ed.), Hakluyt Society, Cambridge, 1971, p. 98.

2 Clements Markham (trans. and ed.), *The Voyages of Pedro Fernández de Quirós, 1595–1606*, Hakluyt Society, Cambridge, 1904, 1967, vol. I, p. 17.

3 The first version appears in Markham, *The Voyages of Pedro Fernández de Quirós*, vol. I, p. 22; the other in *A Collection of Voyages and Travels*, London, 1764, p. 696.

4 *A Collection of Voyages and Travels, some now first printed from original manuscripts, others now first published in English. In six volumes. With a General Preface giving an Account of the Progress of Navigation from its first Beginning*, 3rd edn, vol. 5, A. Churchill and J. Churchill, London, 1746, p. 696. (Fragment evidently written by Suarez de Figueroa.)

5 Markham, *The Voyages of Pedro Fernández de Quirós*, vol. I, p. 21.

6 ibid.

7 ibid., p. 25.

8 ibid., p. 29.

9 Pedro Fernández de Quirós, 'An Account of the Journey', in Antonio de Morga, *Successos de las Islas Filipinas*, J.S. Cummings (trans. and ed.), Hakluyt Society, Cambridge University Press, Cambridge, 1971, pp. 100–1.

10 Markham, *The Voyages of Pedro Fernández de Quirós*, vol. I, pp. 36–7.

11 ibid., p. 39.

12 Roger C. Green, 'The conquest of the conquistadors', *World Archaeology* 5: 1 (June 1973), pp. 14–31, in *Austrialia Franciscana*, VI, Appendix 2, eds Gerard Bushell en colaboración con Celsus Kelly, Franciscan Historical Studies (Australia) and Archivo Ibero-Americano (Madrid), Madrid, 1973, pp. 237–9.

13 Markham, *The Voyages of Pedro Fernández de Quirós*, vol. I, p. 40.

14 ibid., p. 41.

15 Jim Allen and Roger C. Green, 'Mendaña and the fate of the lost "Almiranta", 1595', in *Austrialia Franciscana*, VI, ed. Gerard Bushell in collaboration with Celsus Kelly, Franciscan Historical Studies (Australia) and Archivo Ibero-Americano (Madrid), Madrid, 1973, pp. 235–7.

16 Markham, *The Voyages of Pedro Fernández de Quirós*, vol. I, pp. 62–3.

17 ibid., p. 70.

18 ibid., pp. 72–3.

19 ibid., p. 79.

20 Celsus Kelly, *Calendar of Documents: Spanish Voyages in the South Pacific from Alvaro de Mendaña to Alejandro Malaspina 1567–1794 and the Franciscan Missionary Plan for the Peoples of the Austral Lands 1617–1634*, Franciscan Historical Studies in association with Archivo Ibero-Americano, Madrid, 1965, p. 160.

21 Markham, *The Voyages of Pedro Fernández de Quirós*, vol. I, p. 88.

Chapter 9 Journey to Manila

1 Clements Markham (trans. and ed.), *The Voyages of Pedro Fernández de Quirós, 1595–1606*, Hakluyt Society, Cambridge, 1904, 1967, vol. I, p. 96.

2 ibid., p. 97.

3 ibid., p. 98.

4 ibid., p. 103.

5 ibid., p. 115.

6 Antonio de Morga, *Successos de las Islas Filipinas,* J.S. Cummins (trans. and ed.), Hakluyt Society, Cambridge, 1971, p. 104.

7 Markham, *The Voyages of Pedro Fernández de Quirós*, vol. I, p. 115.

8 *A Collection of Voyages and Travels, some now first printed from original manuscripts, others now first published in English. In six volumes. With a General Preface giving an Account of the Progress of Navigation from its first Beginning*, 3rd edn, A. Churchill and J. Churchill, London, 1746, vol. 5, p. 702.

9 Markham, *The Voyages of Pedro Fernández de Quirós*, vol. I, p. 121.

10 ibid., p. 126. In his letter to Dr Antonio de Morga, Governor of the Philippines, Quirós gave the date as 2 February. See Markham, as above, p. 157.

11 ibid., p. 133.

12 ibid.

13 ibid., p. 146.

CHAPTER 10 QUIRÓS

1 Clements Markham (trans. and ed.), *The Voyages of Pedro Fernández de Quirós, 1595–1606*, Hakluyt Society, Cambridge, 1904, 1967, vol. I, p. 161.

2 Henry N. Stevens (ed.) and George F. Barwick (trans.), 'Relación de don Diego de Prado', in *New Light on the Discovery of Australia as Revealed by the Journal of Captain Don Diego de Prado y Tovar*, Hakluyt Society, London, 1930; Kraus Reprint, Nendeln/Liechtenstein, 1967, p. 241.

3 'mi Rey y Señor natural', in Celsus Kelly, *Pedro Fernándes de Queirós, the Last Great Portuguese Navigator*, Congresso Internacional de História dos Descobrimentos, Lisboa, 1961, p. 7.

4 The Portuguese spelling of his name would have been Fernándes de Queirós.

5 Markham, *The Voyages of Pedro Fernández de Quirós*, vol. I, pp. 294–5.

6 'escribano de una nao de mercadores y portugueses', in Celsus Kelly, *Pedro Fernándes de Queirós*, p. 7.

7 Antonio de Morga, *Successos de las Islas Filipinas*, J.S. Cummins (trans. and ed.), Hakluyt Society, Cambridge, 1971, pp. 104–5.

8 Markham, *The Voyages of Pedro Fernández de Quirós*, vol. I, p. 157.

9 Morga, *Successos*, p. 100.

10 Juan de Iturbe, 'The Sumario Breve de Juan de Iturbe', in *La Austrialia del Espíritu Santo: The Journal of Fray Martín de Munilla O.F.M. and other documents relating to the voyage of Pedro Fernández de Quirós to the South Sea (1605–1606) and the Franciscan Missionary Plan (1617–1627)*, Celsus Kelly (trans. and ed.), vol. II, Hakluyt Society, Cambridge, 1966, p. 276.

11 Celsus Kelly, *Calendar of Documents: Spanish Voyages in the South Pacific from Alvaro de Mendaña to Alejandro Malaspina 1567–1794 and the Franciscan Missionary Plan for the Peoples of the Austral Lands 1617–1634*, Franciscan Historical Studies in association with Archivo Ibero-Americano, Madrid, 1965, p. 174.

12 Markham, *The Voyages of Pedro Fernández de Quirós*, vol. I, p. 162.

13 ibid., p. 162.

14 ibid., p. 163.

15 Celsus Kelly, *Pedro Fernandes de Queirós, the Last Great Portuguese Navigator*, pp. 9–10.

16 Celsus Kelly, in *Some Early Maps Relating to the Queirós–Torres Discoveries of 1606*, Congresso Internacional de História dos Descobrimentos, Lisboa, 1961, dates the *Treatise* to c. 1610.

17 Markham, *The Voyages of Pedro Fernández de Quirós*, vol. I, p. 165.

18 Celsus Kelly, *Calendar of Documents*, p. 188.

19 Markham, *The Voyages of Pedro Fernández de Quirós,* vol. I, pp. 172–3. The Indians were probably the fiercely warlike Caribs, who had replaced the original Arawak inhabitants and against whom Spanish attempts at settlements did not succeed until 1626. Several small islands now constitute the French overseas department of Guadeloupe.

20 Markham, *The Voyages of Pedro Fernández de Quirós*, vol. I, p. 175.

21 Celsus Kelly, *Calendar of Documents*, p. 186.

CHAPTER 11   THE THIRD VOYAGE: 1605–1606

1 Clements Markham (trans. and ed.), *The Voyages of Pedro Fernández de Quirós, 1595–1606*, Hakluyt Society, Cambridge, 1904, 1967, vol. I, p. 177.

2 ibid., p. 178.

3 Celsus Kelly (trans. and ed.), *La Austrialia del Espíritu Santo: The Journal of Fray Martín de Munilla O.F.M. and other documents relating to the voyage of Pedro Fernández de Quirós to the South Sea (1605–1606) and the Franciscan Missionary Plan (1617–1627)*, Hakluyt Society, Cambridge, 1966, vol. II, p. 357.

4 ibid., p. 311.

5 ibid., vol. I, p. 27.

6. ibid., p. 27.

7 Henry N. Stevens (ed.) and George F. Barwick (trans.), *New Light on the Discovery of Australia as revealed by the Journal of Captain Don Diego de Prado y Tovar*, Hakluyt Society, London, 1930; Kraus Reprint, Nendeln/Liechtenstein, 1967, p. 89.

8 Kelly, *La Austrialia del Espíritu Santo*, vol. II, pp. 333–40.

9 ibid., vol. I, Note 6, p. 38.

10 ibid., p. 29. In his narrative Iturbe gives the sum in pesos at over 200 000. As there were several types of peso, it is difficult to make a close comparison with the royal accountant's figure, which is evidently the more precise.

11 Váez also appears as Váes, Báez or Báes.

12 Kelly, *La Austrialia del Espíritu Santo*, vol. I, p. 32.

13 Markham, *The Voyages of Pedro Fernández de Quirós*, vol. I, p. 179.

14 ibid.

15 ibid., pp. 179–80.

16 ibid., p. 181.

Chapter 12 The Voyage Begun

1 Clements Markham (trans. and ed.), *The Voyages of Pedro Fernández de Quirós, 1595–1606*, Hakluyt Society, Cambridge, 1904, 1967, vol. I, p. 182.

2 ibid., vol. II, p. 325.

3 ibid.

4 Celsus Kelly (trans. and ed.), *La Austrialia del Espíritu Santo: The Journal of Fray Martín de Munilla O.F.M. and other documents relating to the voyage of Pedro Fernández de Quirós to the South Sea (1605–1606) and the Franciscan Missionary Plan*

(1617 1627), vol. II, Hakluyt Society, Cambridge, 1966, pp. 144–5.

5 Markham, *The Voyages of Pedro Fernández de Quirós*, vol. I, p. 184.

6 ibid., pp. 184–5.

7 Kelly, *La Austrialia del Espíritu Santo*, vol. II, p. 278.

8 Markham, *The Voyages of Pedro Fernández de Quirós*, vol. I, p. 181.

9 Celsus Kelly, *Calendar of Documents: Spanish Voyages in the South Pacific from Alvaro de Mendaña to Alejandro Malaspina 1567–1794 and the Franciscan Missionary Plan for the Peoples of the Austral Lands 1617–1634*, Franciscan Historical Studies in association with Archivo Ibero-Americano, Madrid, 1965, p. 194.

10 'blanca con una cruz de calatrava en el medio', in Henry N. Stevens (ed.) and George F. Barwick (trans.), *New Light on the Discovery of Australia as Revealed by the Journal of Captain Don Diego de Prado y Tovar*, Hakluyt Society, London, 1930; Kraus Reprint, Nendeln/Liechtenstein, 1967, p. 90.

11 Markham, *The Voyages of Pedro Fernández de Quirós*, vol. I, p. 191.

12 Stevens, *New Light on the Discovery of Australia*, p. 93.

13 Markham, *The Voyages of Pedro Fernández de Quirós*, vol. II, p. 456.

14 ibid., vol. I, pp. 191–3.

15 Celsus Kelly, vol. II, p. 279.

16 Markham, *The Voyages of Pedro Fernández de Quirós*, vol. II, p. 330.

17 Kelly, *La Austrialia del Espíritu Santo*, vol. I, p. 157.

18 Markham, *The Voyages of Pedro Fernández de Quirós*, vol. I, p. 196.

19 ibid., vol. II, p. 333.

20 ibid.

21 Stevens, *New Light on the Discovery of Australia*, p. 101.

22 Markham, *The Voyages of Pedro Fernández de Quirós*, vol. I, p. 197.

23 Kelly, *La Austrialia del Espíritu Santo*, vol. I, p. 163.

24 Markham, *The Voyages of Pedro Fernández de Quirós*, vol. I, p. 205.

25 ibid., vol. II, p. 206.

26 ibid., vol. I, p. 206.

27 ibid., p. 207.

28 ibid., p. 210.

29 ibid., vol. II, p. 345.

30 ibid., pp. 345–6.

31 ibid., vol. I, p. 211.

32 ibid., vol. II, pp. 350–1.

33 ibid.

34 Stevens, *New Light on the Discovery of Australia*, p. 113.

35 Markham, *The Voyages of Pedro Fernández de Quirós*, vol. I, p. 220.

36 Kelly, *La Austrialia del Espíritu Santo*, vol. I, p. 185.

37 ibid., p. 46.

38 Markham, *The Voyages of Pedro Fernández de Quirós*, vol. I, p. 229.

39 ibid., pp. 226–7.

40 Stevens, *New Light on the Discovery of Australia*, p. 223.

41 ibid., p. 113.

CHAPTER 13 LA AUSTRIALIA DEL ESPÍRITU SANTO

1 Clements Markham (trans. and ed.), *The Voyages of Pedro Fernández de Quirós, 1595–1606*, Hakluyt Society, Cambridge, 1904, 1967, vol. I, p. 234.

2 Celsus Kelly (trans. and ed.), *La Austrialia del Espíritu Santo: The Journal of Fray Martín de Munilla O.F.M. and other documents relating to The Voyage of Pedro Fernández de Quirós to the South Sea (1605–1606) and the Franciscan Missionary Plan (1617–1627)*, vol. I, Hakluyt Society, Cambridge, 1966, p. 199.

3 Markham, *The Voyages of Pedro Fernández de Quirós*, vol. I, Note 1, p. 237.

4 Kelly, *La Austrialia del Espíritu Santo*, vol. I, p. 200.

5 Markham, *The Voyages of Pedro Fernández de Quirós*, vol. II, p. 370.

6 ibid., vol. I, p. 240.

7 ibid., vol. II, p. 370.

8 The depth varies somewhat in other accounts.

9 Markham, *The Voyages of Pedro Fernández de Quirós*, vol. I, p. 241.

10 Kelly, *La Austrialia del Espíritu Santo*, vol. I, p. 206.

11 Markham, *The Voyages of Pedro Fernández de Quirós*, vol. II, p. 374.

12 ibid., vol. I, p. 244.

13 Kelly, *La Austrialia del Espíritu Santo*, vol. I, p. 211.

14 Henry N. Stevens (ed.) and George F. Barwick (trans.), *New Light on the Discovery of Australia as Revealed by the Journal of Captain Don Diego de Prado y Tovar*, Hakluyt Society, London, 1930; Kraus Reprint, Nendeln/Liechtenstein, 1967, pp. 122–3.

15 'Letter from Luis Vaez de Torres to the King of Spain', in Markham, *The Voyages of Pedro Fernández de Quirós*, vol. II, p. 461.

CHAPTER 14 KNIGHTHOODS AND A CITY OF MARBLE

1 Celsus Kelly (trans. and ed.), *La Austrialia del Espíritu Santo: The Journal of Fray Martín de Munilla O.F.M. and other docu-

ments relating to *The Voyage of Pedro Fernández de Quirós to the South Sea (1605–1606) and the Franciscan Missionary Plan (1617–1627)*, vol. I, Hakluyt Society, Cambridge, 1966, p. 215.

2 ibid., p. 217.

3 Clements Markham (trans. and ed.), *The Voyages of Pedro Fernández de Quirós, 1595–1606*, Hakluyt Society, Cambridge, 1904, 1967, vol. II, p. 379.

4 ibid., p. 249.

5 The Spanish *repartimiento* or *encomienda* system allotted the labour of conquered people to the conquerors. In return the conquistador was responsible for the natives' Christian instruction and welfare and to provide opportunity to cultivate land, work mines, etc., according to law.

6 Markham, *The Voyages of Pedro Fernández de Quirós*, vol. II, pp. 250–1.

7 ibid., vol. I, p. 253.

8 Kelly, *La Austrialia del Espíritu Santo*, vol. I, p. 286.

9 Henry N. Stevens (ed.) and George F. Barwick (trans.), *New Light on the Discovery of Australia as Revealed by the Journal of Captain Don Diego de Prado y Tovar*, Hakluyt Society, London, 1930; Kraus Reprint, Nendeln/Liechtenstein, 1967, p. 123.

10 Markham, *The Voyages of Pedro Fernández de Quirós*, vol. I, p. 255.

11 ibid., vol. II, p. 385.

12 ibid.

13 ibid.

14 ibid., vol. I, p. 256.

15 ibid., vol. I, p. 261.

16 José Garanger, 'Archaeology of the New Hebrides: contribution to the knowledge of the central islands', Rosemary
Groube (trans.), *Oceania Monograph* 24, 1982.

CHAPTER 15   SEPARATION

1 Celsus Kelly (trans. and ed.), *La Austrialia del Espíritu Santo:
The Journal of Fray Martín de Munilla O.F.M. and other documents relating to The Voyage of Pedro Fernández de Quirós to
the South Sea (1605–1606) and the Franciscan Missionary Plan
(1617–1627)*, vol. II, Hakluyt Society, Cambridge, 1966,
p. 287.

2 Clements Markham (trans. and ed.), *The Voyages of Pedro Fernández de Quirós, 1595–1606*, Hakluyt Society, Cambridge,
1904, 1967, vol. I, p. 280.

3 ibid., vol. II, p. 395.

4 Kelly, *La Austrialia del Espíritu Santo*, vol. II, p. 288.

5 Markham, *The Voyages of Pedro Fernández de Quirós*, vol. II,
p. 396.

6 Kelly, *La Austrialia del Espíritu Santo*, vol. II, p. 264.

7 ibid., p. 300.

CHAPTER 16   THE LAST JOURNEYS OF QUIRÓS

1 Henry N. Stevens (ed.) and George F. Barwick (trans.), *New Light
on the Discovery of Australia as Revealed by the Journal of Captain
Don Diego de Prado y Tovar*, Hakluyt Society, London, 1930; Kraus
Reprint, Nendeln/Liechtenstein, 1967, pp. 211–13.

2 'con esto se tenga por despachado este hombre', in Celsus
Kelly, *Calendar of Documents: Spanish Voyages in the South Pacific
from Alvaro de Mendaña to Alejandro Malaspina 1567–1794 and
the Franciscan Missionary Plan for the Peoples of the Austral Lands*

1617–1634, Franciscan Historical Studies in association with Archivo Ibero-Americano, Madrid, 1965, p. 306.

3 Clements Markham (trans. and ed.), *The Voyages of Pedro Fernández de Quirós, 1595–1606,* Hakluyt Society, Cambridge, 1904, 1967, Appendix VIII, vol. II, p. 525.

Chapter 17 A Different Commander

1 Henry N. Stevens (ed.) and George F. Barwick (trans.), *New Light on the Discovery of Australia as Revealed by the Journal of Captain Don Diego de Prado y Tovar,* Hakluyt Society, London, 1930; Kraus Reprint, Nendeln/Liechtenstein, 1967, p. 229.

2 ibid.

3 ibid., p. 137.

4 ibid., p. 151.

5 Above quotes from Stevens, *New Light on the Discovery of Australia,* pp. 151, 153, 155.

6 Clements Markham (trans. and ed.), *The Voyages of Pedro Fernández de Quirós, 1595–1606,* Hakluyt Society, Cambridge, 1904, 1967, vol. II, p. 463. This translation of Torres' letter was made by Alexander Dalrymple from the original letter at the Archivo General de Simancas in Spain. Some translations made from a copy of the letter omit the words 'There is . . . became . . .'. In Spanish this reads, 'ay por todo el un arcipielago de yslas sin numero por las quales fuimos pesando y al remate de los once grado, yba . . .'.

Chapter 18 The Traverse of Torres Strait

1 A.R. Hinks, 'The discovery of Torres Strait', *The Geographical Journal,* vol. XCVIII, July to December 1941, pp. 91–102.

2 Henry N. Stevens (ed.) and George F. Barwick (trans.), *New Light on the Discovery of Australia as Revealed by the Journal of Captain Don Diego de Prado y Tovar*, Hakluyt Society, London, 1930; Kraus Reprint, Nendeln/Liechtenstein, 1967, p. 231.

3 ibid., p. 165.

4 ibid., p. 153.

5 ibid., p. 161.

6 ibid., pp. 161–3.

7 ibid., p. 163.

8 ibid., p. 165.

9 ibid., p. 239.

10 ibid., p. 165. Torres gives the time on the *placel* as 40 days and in his report to the king as two months.

11 ibid., pp. 231–3.

Chapter 19 'Where are we?'

1 Henry N. Stevens (ed.) and George F. Barwick (trans.), *New Light on the Discovery of Australia as Revealed by the Journal of Captain Don Diego de Prado y Tovar*, Hakluyt Society, London, 1930; Kraus Reprint, Nendeln/Liechtenstein, 1967, p. 233.

2 ibid., pp. 231–3.

3 ibid., p. 230.

4 ibid., p. 175.

5 ibid., p. 177.

6 ibid., p. 233.

7 ibid., p. 181.

8 ibid., p. 235.

Chapter 20 Ternate and Manila

1 Henry N. Stevens (ed.) and George F. Barwick (trans), *New*

Light on the Discovery of Australia as Revealed by the Journal of Captain Don Diego de Prado y Tovar, Hakluyt Society, London, 1930; Kraus Reprint, Nendeln/Liechtenstein, 1967, p. 215.

2 ibid., p. 237.

3 ibid., p. 87.

Chapter 21 A Legacy of Islands

1 Celsus Kelly, *Some Early Maps Relating to the Queirós–Torres Discoveries of 1606*, Congresso Internacional de História dos Descobrimentos, Lisboa, 1961, Note 26, p. 9.

2 Patrick F. Moran, *Discovery of Australia by De Quiros, 1606*, The Catholic Truth Society, Melbourne, 1907.

BIBLIOGRAPHY

PRIMARY SOURCES

A Collection of Voyages and Travel, some now first printed from original manuscripts, others now first published in English. In six volumes. With a General Preface giving an Account of the Progress of Navigations from its first Beginning, 3rd edn, vol. 5, A. Churchill and J. Churchill, London, 1746.

Amherst, William Amhurst Thyssen, Baron, and Thomson, Basil (trans. and eds), *The Discovery of the Solomon Islands by Alvaro de Mendaña in 1568*, vols I and II, Hakluyt Society, London, 1901.

Aristotle, *On the Heavens*, W.K.C. Guthrie (trans.), Heinemann, Cambridge, Massachusetts, 1959.

Bushell, Gerard (ed.), en colaboración con Celsus Kelly, *Austrialia Franciscana, Documentos sobre la expedición de Alvaro de Mendaña para poblar las Islas de Salomón (1595–1597), Relaciones de la población de las Islas de Santa Cruz, el fracaso y sus consecuencías inmediatas*, vol. VI, Franciscan Historical Studies (Australia) y el Archivo Ibero-Americano (Madrid), Madrid, 1973.

Camoens, Luis Vaz de, *The Luciads*, William C. Atkinson (trans.), Penguin Books, Harmondsworth, 1952.

Costa, H. de la (ed.), *Readings in Philippine History: Selected Historical Texts Presented with a Commentary*, Bookmark, Manila, 1955.

Kelly, Celsus, *Austrialia Franciscana*, vol. I, *Documentos Franciscanos sobre la expedición de Pedro Fernández de Quirós al Mar del Sur (1605–1606), y planes misionales sobre la conversión de los nativos de las Tierras Australes (1617–1634)*, Franciscan Historical Studies (Australia) en colaboración con Archivo Ibero-Americano (Madrid), Madrid, 1963.

——*Austrialia Franciscana*, vol. II, *Relaciones de Alvaro de Mendaña al Rey Don Felipe II y de Gómez Hernández Catoira al Gobernador del Perú, Don Lope García de Castro, sobre la expedición de Mendaña a las Islas de Salomón en el Mar del Sur (1567–1569)*, Franciscan Historical Studies (Australia) en colaboración con Archivo Ibero-Americano (Madrid), Madrid, 1965.

——*Austrialia Franciscana*, vol. III, *Documentos sobre la expedición de Alvaro de Mendaña a las Islas de Salomón en el Mar del Sur (1567–1569), Preparativos para la expedición y Relaciones sobre la misma*, Franciscan Historical Studies (Australia) en colaboración con Archivo Ibero-Americano (Madrid), Madrid, 1967.

——*Austrialia Franciscana*, vol. IV, *Documentos sobre la expedición de Alvaro de Mendaña a las Islas de Salomón en el Mar del Sur (1567–1569), Relaciones de la expedición y una selección de los documentos posteriores sobre la misma,* Franciscan Historical Studies (Australia) en colaboración con Archivo Ibero-Americano (Madrid), Madrid, 1969.

———*Austrialia Franciscana*, vol.V, *Una selección de noventa documentos sobre las negociaciones de Alvaro de Mendaña en la Corte de Madrid, en Panamá y el Perú para poblar las Islas de Salomón (1572–1595)*, Franciscan Historical Studies (Australia) en colaboración con Archivo Ibero-Americano (Madrid), Madrid, 1971.

———*La Austrialia del Espíritu Santo: The Journal of Fray Martín de Munilla O.F.M. and other documents relating to The Voyage of Pedro Fernández de Quirós to the South Sea (1605–1606) and the Franciscan Missionary Plan (1617–1627)*, vols I and II, Hakluyt Society, Cambridge, 1966.

Major, R.H. (ed.), *Early Voyages to Terra Australis, now called Australia: A Collection of Documents and Extracts from Early Manuscript Maps, Illustrative of the History of Discovery on the Coasts of that Vast Land, from the Beginning of the Sixteenth Century to the Time of Captain Cook*, Hakluyt Society, London, 1859.

Markham, Clements (trans. and ed.), *The Voyages of Pedro Fernández de Quirós, 1595 to 1606*, vols I and II, Hakluyt Society, London, 1904; Kraus Reprint Ltd, Nendeln/Liechtenstein, 1967.

Middleton, Sir Henry, *The Voyage of Sir Henry Middleton to the Moluccas 1604–1606*, Hakluyt Society, London, 1943.

Pigafetta, Antonio, *Magellan's Voyage: A Narrative Account of the First Navigation*, R. A. Skelton (trans.), The Folio Society, London, 1975.

Queirós, Pedro Fernándes de, *Descubrimiento de la regiones austriales*, Roberto Ferrando (ed.), Historia 16, Información y Revistas, SA, Madrid, 1986.

Quirós, Pedro Fernández, *Descubrimiento de Austrialia, Memorial No. 8, Texto original (impreso en Madrid c. 1609) y presentado al Rey Felipe III por el Capitán Pedro Fernández de Quirós*, Carlos Sanz, Madrid, 1962.

——*Memoriales de las Indias Australes*, Oscar Pinochet (ed.), Historia 16, Información y Revistas, SA, Madrid, 1991.

——*Memorial n. 8 presentado al Rey Felipe III por el Capitán Pedro Fernández de Quirós, sobre la Población y Descubrimiento de la 'Quarta Parte del Mundo' Austrialia Incognita*, Carlos Sanz, Biblioteca Austrialiana Vertustísima, 1964.

SECONDARY SOURCES

Alexander, James M. *The Islands of the Pacific, from the Old to the New*, American Tract Society, New York, 1895.

Allen, Paul C., *Philip III and the Pax Hispanica 1598–1621: The Failure of the Grand Strategy*, Yale University Press, New Haven and London, 2000.

Australian Dictionary of Biography, 1788–1850, vol. 2, Melbourne University Press, Melbourne, 1966.

Badger, Geoffrey, *The Explorers of the Pacific*, Kangaroo Press, Sydney, 1988 and 1996.

Bagrow, Leo, *History of Cartography*, rev. by R.A. Skelton, C.A. Watts, London, 1964.

Barclay, Glen, *A History of the Pacific from the Stone Age to the Present Day*, Sidgwick & Jackson, London, 1978.

Barrett, Charles, *The Pacific Ocean of Islands*, N.H. Seward, Melbourne, 1950.

Beaglehole, J.C., *The Exploration of the Pacific*, Stanford University Press, Stanford, CA, 1966.

Bergreen, Laurence, *Over the Edge of the World: Magellan's Terrifying Circumnavigation of the Globe*, Harper Perennial, London, 2003.

Bertrand, Louis, *The History of Spain, Part I: from the Visigoths to the Death of Philip II*, Dawsons of Pall Mall, London, 1969.

Black, Jeremy, *Visions of the World*, Mitchell Beazley, London, 2003.

Blair, Emma Helen, and Robertson, James Alexander, *The Philippine Islands, 1493–1803*, vols II, XIII and XIV, A.H. Clark, Cleveland, 1903–1909.

Brownlee, Walter, *The First Ships Round the World*, Cambridge University Press, Cambridge, 1974.

Bunbury, E.H., *A History of Ancient Geography Among the Greeks and Romans from the Earliest Ages Till the Fall of the Roman Empire*, vol. I, Dover Publications, New York, 1959.

Burnett, Pat, *Rediscovering Australia*, published by the author, 1996.

Burney, James, *A Chronological History of the Discoveries in the South Sea or Pacific Ocean*, Parts I and II, Luke Hansard, London, 1803–1817.

Callander, John, *Terra Australis Cognita or, Voyages to the Terra Australis or Southern Hemisphere, during the Sixteenth, Seventeenth, and Eighteenth Centuries*, Hawes, Clark & Collins, London, 1768; Facsimile, N. Israel, Amsterdam, 1967.

Carr, Raymond, *Spain: A History*, Oxford University Press, Oxford, 2000.

Chirino, Pedro, *Relación de las Islas Filipinas—The Philippines in 1600*, Ramón Echevarria (trans.), Historical Conservation Society, Manila, 1969.

Clark, C.M.H., *A History of Australia, from the Earliest Times to the Age of Macquarie*, Melbourne University Press, Melbourne, 1962.

Collingridge, George, *The Discovery of Australia: A Critical, Documentary and Historic Investigation Concerning the Priority of Discovery in Australasia by Europeans before the Arrival of Lieut. James Cook, in the 'Endeavour' in the year 1770*, Hayes

Brothers, Sydney, 1895; Golden Press Facsimile Edition, Sydney, 1983.

Coote, Stephen, *Drake, The Life and Legend of an Elizabethan Hero*, Simon & Schuster, Sydney, 2003.

Crone, G.R., *Maps and Their Makers: An Introduction to the History of Cartography*, Hutchinson University Library, London, 1966.

Crow, John A., *Spain: The Root and the Flower*, University of California Press, Berkeley, 1985.

Cushner, Nicholas P., *Spain in the Philippines, from Conquest to Revolution*, Ateneo de Manila University, Quezon City, 1971.

Dalrymple, Alexander, *An Account of the Discoveries made in the South Pacifick Ocean*, Hordern House Rare Books, Sydney, 1996.

——*An Historical Collection of the Several Voyages and Discoveries in the South Pacific Ocean*, vol. I, J. Nourse et al, London, 1770–1.

——*Chart of the South Pacific Ocean, pointing out the discoveries made therein previous to 1764*, printed for the author, London, 1770.

Dalton, Bill, *A Traveller's Notes: Indonesia* (Cover title: *Indonesia and Papua New Guinea*), Moon Publications, Melbourne, 1976.

Dunmore, John, *Who's Who in Pacific Navigation*, Melbourne University Press, Melbourne, 1992.

Dunn, F.M., *Quiros Memorials: A Catalogue of Memorials by Pedro Fernandez de Quiros—1607–1615, in the Dixson and Mitchell Libraries*, Sydney, 1961.

Ellis, William, *Polynesian Researches, during a Residency of Nearly Six Years in the South Sea Islands; including descriptions of the natural history and scenery of the islands—with remarks on the history, mythology, traditions, government, arts, manners and customs of the inhabitants*, vol. I, Dawsons of Pall Mall, London, 1967.

Emery, James, *The Discovery of Australia, Including the Mandated Territory of New Guinea*, Hamlyn Guies, Sydney, 1973.

Estensen, Miriam, *Discovery: The Quest for the Great South Land*, Allen & Unwin, Sydney, 1998.

Feeken, Erwin H.J. et al, *The Discovery and Exploration of Australia*, Thomas Nelson, Melbourne, 1970.

Fernández-Shaw, Carlos M., *España y Australia, Cinco Siglos de Historia (Spain and Australia, Five Centuries of History)*, Edición Alonso Ibarrola y Mercedes Palau, Dirección General de Relaciones Culturales y Científicas, Ministerio de Asuntos Exteriores de España, 2000.

——*España y Australia, Quinientos años de relaciones*, Dirección General de Relaciones Culturales y Científicas, Ministerio de Asuntos Exteriores de España, 2001.

Ferrando, Roberto, *Descubrimiento de las regiones australes*, Historia 16, Información y Revistas, SA, Madrid, 1986.

Ferrer de Couto, José, *Historia de la Marinha Real Española: desde el descubrimiento de las Américas hasta el combate de Trafalgar*, por José March y Labores, J.M. Ducazcal, Madrid 1854.

Findlay, Alexander George, *Directory for the Navigation of the South Pacific Ocean, with Descriptions of the Coasts, Islands, etc., from the Strait of Magalhaens to Panama, and those of New Zealand, Australia, etc., its Winds, Currents, and Passages*, Richard Homes Laurie, London, 1863.

Fitzgerald, Lawrence, *Java la Grande: The Portuguese Discovery of Australia*, The Publishers, Hobart, 1984.

Garvey, Robert, *To Build a Ship: The VOC Replica Ship, Duyfken*, University of Western Australia Press, Perth, 2001.

Gash, Noel, and Whittaker, June, *A Pictorial History of New Guinea*, Robert Brown & Associates, Brisbane, 1975.

Gibbons, Tony (general ed.), *The Encyclopedia of Ships*, Silverdale Books, Enderby, Leicester, 2001.

Gibson, Charles, *Spain in America*, Harper Colophon Books, New York, 1966.

Gill, J.C.H., *The Missing Coast: Queensland Takes Shape*, Queensland Museum, Brisbane, 1988.

Graham, Winston, *The Spanish Armadas*, Penguin Books, Harmondsworth, 1972.

Hanson, Neil, *The Confident Hope of a Miracle: the True History of the Spanish Armada*, Doubleday, London, 2003.

Harcombe, David, and Honan, Mark, *Solomon Islands*, 3rd edn, Lonely Planet, Melbourne, 1997.

Hardy, John, and Frost, Alan (eds), *Studies from Terra Australis to Australia*, Australian Academy of the Humanities, Canberra, 1989.

Hawthorne, Daniel, *Islands of the Pacific*, G.P. Putnam's Sons, New York, 1943.

Henderson, James, *Sent Forth a Dove: Discovery of the Duyfken*, University of Western Australia Press, Perth, 1999.

Hezel, Francis X., *The First Taint of Civilization: A History of the Caroline and Marshall Islands in Pre-Colonial Days, 1521–1885*, University of Hawaii Press, Honolulu, 1983.

Hilder, Brett, *El viaje de Torres de Vera Cruz a Manila: Descubrimiento de la costa meridional de Nueva Guinea y del Estrecho de Torres y Documentos de la época de la travesía*, Franciso Utray (ed.), Ministeria de Asuntos Exteriores, Madrid, 1990.

——*The Voyage of Torres: Discovery of the Southern Coastline of New Guinea and Torres Strait by Captain Luis Baéz de Torres in 1606*, University of Queensland Press, Brisbane, 1980.

Howgego, Raymond John, *Encyclopaedia of Exploration to 1800*, Hordern House, Sydney, 2003.

Jack-Hinton, Colin, *The Search for the Islands of Solomon, 1567–1838*, Clarendon Press, Oxford, 1969.

Kamen, Henry, *Spain's Road to Empire: the Making of a World Power 1492–1763*, Penguin Press, London, 2002.

Kelly, Celsus, *Calendar of Documents: Spanish Voyages in the South Pacific from Alvaro de Mendaña to Alejandro Malaspina 1567–1794 and the Franciscan Missionary Plan for the Peoples of the Austral Lands 1617–1634*, Franciscan Historical Studies (Australia) in association with Archiva Ibero-Americano (Madrid), Madrid, 1965.

——*Pedro Fernandes de Quierós, the Last Great Portuguese Navigator*, Congresso Internacional de Historia dos Descubrimentos, Lisbon, 1961.

——*Some Early Maps Relating to the Queirós-Torres Discoveries of 1606*, Congresso Internacional de Historia dos Descubrimentos, Lisboa, 1961.

King, David, and Ranck, Stephen (eds), *Papua New Guinea Atlas: A Nation in Transition*, Robert Brown & Associates (Australia), Bathurst, NSW, 1982.

King, Robert J., *The Secret History of the Convict Colony: Alexandro Malaspina's report on the British settlement of New South Wales*, Allen & Unwin, Sydney, 1990.

Kiple, Kenneth F. (ed.), *Plague, Pox and Pestilence: Disease in History*, Phoenix Illustrated, London, 1999.

Kish, George (ed.), *A Source Book in Geography*, Harvard University Press, Cambridge, Mass., 1978.

Langdon, Robert, *The Lost Caravel*, Pacific Publications, Sydney, 1975.

——*The Lost Caravel Re-Explored*, Brolga Press, Canberra, 1988.

Lanyon, Anna, *The New World of Martin Cortes*, Allen & Unwin, Sydney, 2003.

Lapa, Albino, *Pedro Fernándes de Queirós, o último navegador portugués que descubriu, no ano de 1606, as Ilhas do Espírtu Santo, Nova Hébridas*, Divisão de Publicações e Biblioteca Agência Geral do Ultramar, 1951.

Lee, Stephen J., *Aspects of European History 1494–1789*, 2nd edn, Routledge, London, 1990.

Lopez, Rafael, and Felix, Alfonso (eds), *The Christianization of the Philippines*, Historical Conservation Society and University of San Augustín, Manila, 1965.

Martín, Luis, *Daughters of the Conquistadors:Women of the Viceroyalty of Peru*, University of New Mexico Press, Albuquerque, 1983.

McCarthy, Edward J., *Spanish Beginnings in the Philippines, 1564–1572*, The Catholic University of America Press, Washington, DC, 1943.

MacDonald, Barrie, *Cinderellas of the Empire:Towards a History of Kiribati and Tuvalu*, ANU Press, Canberra, 1982.

Macknight, C.C. (ed.), *The Farthest Coast:A Selection of Writings Relating to the History of the Northern Coast of Australia*, Melbourne University Press, Melbourne, 1969.

Merriman, Roger Bigelow, *The Rise of the Spanish Empire in the Old World and in the New*, vol. 4, Cooper Square, New York, 1962.

Ministro de Asuntos Exteriores, *Spanish Pacific from Magellan to Malaspina*, Comisión del Centenario del Descubrimiento de América, Madrid, 1988.

Moran, Patrick Francis, *Discovery of Australia by De Quiros 1606*, The Catholic Truth Society, Melbourne, 1907.

Morga, Antonio de, *Successos de las Islas Filipinas*, J.S. Cummings (ed. and trans.), Hakluyt Society, Cambridge University Press, Cambridge, 1971.

Mutch, T.D., *The First Discovery of Australia With an Account of the Voyage of the 'Duyfken' and the Career of Captain Willem Jansz*, privately printed, Sydney, 1942.

Nicholson, Ian, *Via Torres Strait: A Maritime History of the Torres Strait Route and the Ships' Post Office at Booby Island*, published by the author, 1996.

Oliver, Douglas, *The Pacific Islands*, Harvard University Press, Cambridge, Mass., 1962.

Olson, James S. (editor-in-chief), *Historical Dictionary of the Spanish Empire*, Greenwood Press, Westport, Conn., 1992.

Parker, Geoffrey, *The Dutch Revolt*, Penguin Press, London, 1977, 1985.

Pérez-Malláina, Pablo E., *Spain's Men of the Sea: Daily Life on the Indies Fleets in the Sixteenth Century*, Carla Rahn Philips (trans.), John Hopkins University Press, Baltimore, 1998.

Perry, T.M., *The Discovery of Australia: The Charts and Maps of the Navigators and Explorers*, Nelson, Melbourne, 1982.

Petrie, Charles, *The History of Spain, Part II: from the death of Philip II to 1945*, Dawsons of Pall Mall, London, 1969.

Pinochet de la Barra, Oscar, *Quirós y su utopía de las Indias Australes*, Ediciones de Cultura Hispánica, Madrid, 1989.

Powell, Alan, *Far Country: a short history of the Northern Territory*, 4th edn, Melbourne University Press, Melbourne, 2000.

Price, A. Grenfell, *Island Continent: Aspects of the Historical Geography of Australia and Its Territories*, Angus & Robertson, Sydney, 1972.

Raisz, Erwin, *Mapping the World*, Abelard-Shumann, New York, 1956.

Reed, Robert R., *Colonial Manila: The Context of Hispanic Urbanism and Process of Morphogenesis*, University of California Press, Berkeley, 1978.

Richardson, W.A.R., *The Portuguese Discovery of Australia: Fact or Fiction?*, National Library of Australia, Canberra, 1989.

Roberts, J.M., *The Penguin History of the World*, Penguin Books, Harmondsworth, 1976, 1997.

Ross, John (editor-in-chief), *Chronicle of Australia*, Penguin, Ringwood, Victoria, 2000.

Sanvitores, Diego Luys de, *Memorial que el P. Diego Luis de Sanvitores: religioso de la Compañâia de Jesâus, Rector de las islas Marianas, remitiâo âa la congregaciâon del glorioso apâostal de las Indias, S Francisco Xavier de la Ciudad de Mâexico, pidiendole ayuden y soccoran para la fundacion de la missiâon de dichas islas*, Francisco Rodrâiguez Lupercio, Mexico, 1669.

Sanz, Carlos, *Australia, its Discovery and Name, with Facsimile reproductions of the Quirós Memorial and other rare illustrations*, Dirección General de Relaciones Culturales, Madrid, 1964.

Sarton, George, *A History of Science: Ancient Science Through the Golden Age of Greece*, W.W. Norton, New York, 1970.

Scarr, Deryck, *The History of the Pacific Islands: Kingdoms of the Reef*, Macmillan, Melbourne, 1990.

Schurz, Willian Lytle, *The Manila Galleon*, E.P. Dutton, New York, 1959.

Scott, William Henry, *Cracks in the Parchment Curtain and Other Essays in Philippine History*, New Day Publishers, Quezon City, 1985.

Sharp, Andrew, *The Discovery of the Pacific Islands*, Oxford University Press, Oxford, 1960.

Sigmond, J.P., and Zuiderbaan, L.H., *Dutch Discoveries of Australia:*

Shipwrecks, Treasures and Early Voyages off the West Coast, Rigby, Adelaide, 1979.

Smith, Roger C., *Vanguard of the Empire: Ships of Exploration in the Age of Columbus*, Oxford University Press, New York, Oxford, 1993.

Spate, O.H.K., *Australia, New Zealand and the Pacific*, vol. I, Oxford University Press, London, 1979.

——*The Pacific Since Magellan*, vol. I, *The Spanish Lake*, ANU Press, Canberra, 1979.

——*The Pacific Since Magellan*, vol. II, *Monopolists and Freebooters*, ANU Press, Canberra, 1983.

Taylor, Andrew, *The World of Gerard Mercator: The Mapmaker who Revolutionised Geography*, Harper Collins, London, 2004.

Thomas, Hugh, *Rivers of Gold: The Rise of the Spanish Empire*, Weidenfeld & Nicolson, London, 2003.

Thomson, J. Oliver, *History of Ancient Geography*, Biblo & Tannen, New York, 1965.

Times Atlas of World Exploration, The, Times Books, London, 1991.

Tooley, R.V., *The Mapping of Australia*, Holland Press, London, 1979.

——*Tooley's Dictionary of Mapmakers*, Map Collector Publications, Ltd, Tring, Hertfordshire, 1979.

Vlekke, Bernard H.M., *Nusantara: A History of the East Indian Archipelago*, Harvard University Press, Cambridge, Mass., 1944.

Wheatcroft, Andrew, *The Habsburgs: Embodying Empire*, Penguin Books, London, 1995, 1996.

Williams, Glyndwr, and Frost, Alan (eds), *Terra Australis to Australia*, Oxford University Press in association with the Australian Academy of the Humanities, Melbourne, 1988.

Wintle, Justin, *The Rough Guide History of Spain*, Rough Guides, London, 2003.

Wood, G. Arnold, *The Discovery of Australia*, Macmillan, London, 1922.

Wroth, Lawrence C., *The Early Cartography of the Pacific* [Washington: s.n.], 1944.

Zaide, Gregorio F., *Philippine Political and Cultural History: The Philippines since Pre-Spanish Times*, vol. I, rev. edn, Philippine Education, Manila, 1957.

Zainu'ddin, Aisa, *A Short History of Indonesia*, Cassel Australia, Melbourne, 1968.

Zaragoza, Ramón Ma, *Old Manila*, Oxford University Press, Oxford, 1990.

ARTICLES

Aurousseau, M., 'Where did Torres pass?', *Newsletter of the Royal Australian Society*, October 1973, pp. 3, 6.

Bayldon, Francis J., 'Voyage of Luis Vaez de Torres from the New Hebrides to the Moluccas, June to November, 1606', *The Royal Australian Historical Society Journal and Proceedings*, vol. XI, Part III, 1925, pp. 158–94.

——'Voyage of Torres', *Royal Australian Historical Society Journal and Proceedings*, vol. XVI, 1930, pp. 133–46.

——'Notes and comments on "New light on the discovery of Australia"', *Royal Australian Historical Society Journal and Proceedings*, vol. XVII, Part V, 1932, pp. 289–330.

——'Voyage of Torres', Corrigenda, *Royal Australian Historical Society Journal and Proceedings*, vol. XVIII, Part I, 1932, pp. 47–8.

——'Remarks on navigators of the Pacific, from Magellan to

Cook', *Royal Australian Historical Society Journal and Proceedings*, vol. XVIII, Part III, 1932.

Clarkson, William, 'Note on Prado's "Relacion"', *Royal Australian Historical Society Journal and Proceedings*, vol. XVI, Supplement to Part II, 1930, pp. 147–50.

Dixson, William, 'Notes and comments on "New light on the discovery of Australia"', *Australian Historical Society Journal and Proceedings*, vol. XVIII, Part V, 1932, pp. 289–329.

Garanger, José, 'Archaeology of the New Hebrides: contribution to the knowledge of the central islands', Rosemary Groube (trans.), *Oceania Monograph* 24, 1982.

Hinks, A.R., 'The discovery of Torres Strait', *The Geographical Journal*, vol. XCVIII, July to December, The Royal Geographical Society, London, 1941, pp. 91–102.

Illidge, P., et al, 'An assessment of selected shipwrecks in Torres Strait and far north Great Barrier Reef', *Memoirs of the Queensland Museum*, Cultural Heritage Series 3 (i): 93–104, Brisbane, 2004.

Lesson, M., 'Notice bibliographique sur Quiros', *Bulletin de la Sociéte de Géographie de Rochefort*, 1883–1884, pp. 25–6.

MacDonald, A.C., 'Alleged Discovery of Australia by de Quiros in 1606: a reply to Cardinal Moran's second pamphlet', *Victorian Geographical Journal*, vol. XXVII, Royal Geographical Society of Australia (Victoria), 1909, pp. 74–88.

McMahon, M.M., 'Great Spanish discoverers: the last of them: Fernandez de Quiros', *The Evening News*, Sydney, 17 August 1910.

O'Brien, Eris M., 'The Basilian monk Don Diego de Prado y Tovar; did he discover Australia in 1606?', *The Australasian Catholic Record*, vol. VII, no. 4, October 1930, pp. 302–19.

Panton, J.A., 'A brief review of the additional evidence in support of the theory that the eastern coast of Australia was discovered by the de Quiros Expedition', *Victorian Geographical Journal*, vol. XXVII, Royal Geographical Society (Victoria), 1909, pp. 66–73.

Parkyn, E.A., 'The voyage of Luis Vaez de Torres: new light on the discovery of Australia, as revealed by the Journal of Don Diego de Prado y Tovar', Henry N. Stevens (ed.), *The Geographical Journal*, vol. LXXVI, The Royal Geographical Society, July–December 1930, London, pp. 252–6.

MISCELLANEOUS

McAuley, James, *Captain Quiros—A Poem*, Angus & Robertson, Sydney, 1964.

Toohey, John, *Quiros*, Duffy & Snellgrove, Sydney, 2002 (a novel).

INDEX

capitana is separated from the other
vessels 175–6; conflicting accounts
of navigation decisions made 176–8;
Quirós' order to proceed towards
Santa Cruz replaced by order to
proceed to Mexico; *capitana* sails to
Mexico 178–81. *See also* Quirós.

voyage (1606–1607) of the *San Pedro*
and launch *Los Tres Reyes* under the
command of Luis Váes de Torres
(after separation from the flagship
San Pedro y San Pablo) chapters
17–20; Torres takes command
187–8; food/food sources and
shortages 188; Torres confirms
that La Austrialia del Espíritu
Santo is an island 188; arrives at
Louisiade Archipelago, follows
the chain to the bay Puerto de
San Francisco 190–1; excursion
ashore to unfriendly reception,
some islanders shot and killed
191–2; Torres takes possession in
the name of Philip III 192; follow
coast to Bona Bona island and
friendly reception, food and water
obtained 193; to San Bartolomeo
(Mailu), conflict with islanders and
children taken 194–5; anchor off
Port Moresby, then through the
Gulf of Papua aiming for Manila
195; enter Torres Strait 196;
anchor off Isla de los Perros (Isle
of Dogs), women taken on board
198–9, 204; pass many islands
199–202; vessels emerge from
Torres Strait and follow coastline
of Irian Jaya 203–4, 205; anchor at
island of Lakahia and then island of
Aiduma, claimed for Philip III; to
McClure Gulf (Teluk Berau) 206;
between Kipulauan Fam islands
and Dampier Strait, a Portuguese
man who comes aboard confirms
the vessels' position as five days

sail from the island kingdom of
Bachan, Moluccas, and tells of a
Spanish garrison on the islands of
Ternate and Tidore 206–7; cross
the Halmahera Sea to Bachan
207–9; the king of Bachan comes
aboard 209–10; subdue rebels at
his request 210–11; sail to Ternate
island and Spanish garrison 213–14;
depart from Ternate, leaving
launch and 20 men 214; enter
harbour of Cavite in Manila Bay
and learn of the arrival of Quirós
and the *capitana* in Mexico 215–16;
Torres plans to refit and reprovision
the *San Pedro* in Manila, Audiencia
refuses permission and the vessel
and crew taken 216–18; no further
record of Torres 218